Counter-Intelligence

Counter-Intelligence
What the Secret Wrld Can Teach Us Ab ut Pr blem-s lving and Creativity
Robert Hannigan

HarperCollins*Publishers*

HarperCollins*Publishers*
1 London Bridge Street
London SE1 9GF

www.harpercollins.co.uk

HarperCollins*Publishers*
Macken House, 39/40 Mayor Street Upper
Dublin 1, D01 C9W8, Ireland

First published by HarperCollins 2024

1 3 5 7 9 10 8 6 4 2

© Robert Hannigan 2024

Robert Hannigan asserts the moral right to
be identified as the author of this work

A catalogue record of this book is
available from the British Library

HB ISBN 978-0-00-839855-2
PB ISBN 978-0-00-839856-9

Printed and bound in the UK using 100%
renewable electricity at CPI Group (UK) Ltd

This book contains FSC™ certified paper and other controlled
sources to ensure responsible forest management.

For more information visit: www.harpercollins.co.uk/green

For the people of Bletchley Park and GCHQ
– past, present and future

Contents

Author's note

I owe many people thanks for helping me to write this book. I am not a historian and have relied very heavily on the work of published experts on Bletchley Park, GCHQ and codebreaking, not least Michael Smith, David Kenyon, Dermot Turing, David Kahn, Andrew Hodges, Jack Copeland, Simon Singh and John Ferris. David Abrutat, GCHQ's expert historian, reviewed the manuscript and made suggestions, though responsibility for the opinions expressed and any errors remains mine. I have tried to acknowledge sources wherever possible and have listed key works in the bibliography. I will put right any omissions in future editions.

Being a Trustee of Bletchley Park has been a privilege and a wonderful excuse to spend time there, and I greatly admire what Iain Standen and his team have achieved in restoring the Park and presenting it to new generations of visitors. The author's profits go to the Bletchley Park Trust.

I am very grateful to those who gave interviews, including Vint Cerf, Victoria Coren Mitchell and Roger Penrose, as well as to Sharon White and the archivists of the John Lewis Partnership for their assistance.

Without necessarily realising it, many colleagues in the secret world and the Northern Ireland Office have inspired parts of this book, including good friends in the United States signals intelligence community. For the avoidance of doubt, I should explain that where I have referred to living staff and given examples or stories, notably on neurodiversity, these are 'fiction- alised': composites of people, places and quotations. That is a polite way of saying I have made them up, partly to maintain confidentiality but also to avoid using real individuals as speci- mens. Nor are there any new revelations here; I hope those fascinated by reading secrets will instead enjoy investigating why secrecy itself is so fascinating to so many.

I also owe a great deal to the mathematicians and 'cryppies' of GCHQ, the heart of the operation from Bletchley Park onwards and, of course, the original puzzlers. This is not a tech- nical book and they will be shocked at my cavalier use of the words 'codes' and 'codebreaking' to cover cryptography, crypta- nalysis, cryptology, ciphers and other areas of their work.

While writing these stories of problem-solving and creativity in technology, it has also been an educational adventure to be part of a cyber security start-up, and I owe Jim Rosenthal and my extraordinarily talented colleagues at BlueVoyant a great debt for allowing me to observe an innovative team doing extraordinary technical things to meet a huge global challenge.

Ursula Martin read the manuscript with an academic's foren- sic perspective and kindly shared her deep knowledge of the history of computer science. Tony Comer, lifelong siginter and GCHQ's first publicly avowed historian, was typically generous with his time and wise insights. More than anyone, Tony has been a custodian and advocate of the culture and history of which I have only touched the surface here.

This book would never have been finished without the ever-patient professionalism of Joel Simons and his colleagues at HarperCollins. Neil Blair originally encouraged me to pursue the idea, and I am hugely grateful to Rory Scarfe at The Blair Partnership for his endless expertise, guidance and good humour over the past few years.

The longer I spend in the academic world, the more historical connections I find to the secret world. I am writing this in a study once occupied by the philosopher Stuart Hampshire, former warden of Wadham College, Oxford. After serving in intelligence during the Second World War, he was commissioned to write a report on what should happen to GCHQ. It is largely thanks to him that the organisation prospered financially under a government that had little understanding of what it did or what the new, costly world of computers might bring. Hampshire understood the centrality of the relationship with the US National Security Agency and how the balance of power was shifting. I am grateful to the Fellows and staff of Wadham for giving me space in which to write.

Finally, Celia, Luke and Evie put up with years of secrecy, starting with an unusual family life in Northern Ireland, followed by recent years of procrastination in writing this. They have maintained a healthy detachment, balance and sense of humour towards the secret world, attitudes that would be very familiar to the best of those who worked at Bletchley Park.

Introduction

Not long after my arrival at GCHQ I met with the senior leadership and we talked about the organisation's strange history. It was a small group, since GCHQ is a 'flat' organisation and one of the few in government where it is relatively difficult to persuade people to go for promotion to leadership roles. Its iconic main building, known as the 'Doughnut', is brilliantly designed, but without much interest in large meeting spaces, for cultural reasons that I hope will become obvious. As a result, we met in an uninspiring concrete-walled room underground to talk about the past – and also the future.

I brought up the subject of Bletchley Park that day, partly because I had studied it for some years and it ran through the DNA of GCHQ – in fact some of our new recruits had trained there until the buildings became too dilapidated in the 1980s. But the main reason I wanted to discuss it was because Bletchley, and some of the achievements of our predecessors in the 1960s and 1970s, seemed to me to have lessons for our approach to the huge new challenges that lay ahead of us.

In 2014 the UK was facing a rising tide of cyber attacks across the economy, the re-emergence of Islamist terrorism in the shape of ISIS, a newly aggressive Russian regime and a re-emergent, assertive China. The dominance of big tech companies was presenting new problems in making high-grade encryption available to bad people as well as good, and increasingly the intelligence picture the West had been used to was 'going dark'.

GCHQ itself had come though the Edward Snowden revelations of the previous year relatively unscathed, albeit subject to a level of publicity that had been deeply uncomfortable for staff and presenting a huge diversion of effort. It now needed to build and secure public trust, and the prime minister at the time was keen to make the organisation more open.

As always, GCHQ had to confront an overabundance of threats with limited resources, augmented by some strategic challenges as the centre of gravity for technology continued to move away from government to the private sector. In the cyber world and the emerging landscape of artificial intelligence, this was a sector spending trillions of dollars each year on research and development, against which GCHQ's budget, even combined with that of its US allies, seemed puny. But government looked to GCHQ to solve hard technical security problems, both in the present and ten or more years ahead. The organisation's task remained to meet present threats and crises while building for the future, to keep re-engineering itself in flight.

In short, GCHQ was in the process of facing the challenges of the age of digital information and intelligent machines, an age that its Bletchley predecessors had done so much to create. This book tries to trace the human journey of an organisation that is always on the brink of irrelevance, of disruption by new adversaries deploying new technologies, always about to be made obsolete by the march of science. Fast-forward to the post-pandemic world, and we have all become acutely aware of our current failures to plan, innovate, manufacture and organise for a nation at scale and at speed.

These questions become more urgent as we survey the global problems piling up ahead of us, from future pandemics to climate change. After a century of dizzyingly rapid technological change, we now need to move faster than ever to create a sustainable future. Problem-solving has become existential for humanity.

Our discussion in the windowless room of the Doughnut was about a thread of innovation and performance in the secret world running through the period around the Second World War. The greatest technological innovation of wartime was the development of programmable computing and its practical use at Bletchley Park to break enemy codes and read their communications. Here was arguably the world's first big data processing and knowledge-management organisation. George Steiner, looking back fifty years after the conflict had ended, described Bletchley as 'the single greatest achievement of Britain during 1939–45, perhaps during the century as a whole'. Much has been written about the contribution of these women and men (three-quarters of the staff were female); as they themselves were the first to point out, they were not on the front line fighting and they never claimed to have won the war, but there is no question that their efforts shortened its course and helped ensure Allied victory. But the legacy of programmable computing dwarfed even this achievement.

The core of Bletchley continued as GCHQ into the Cold War and led to a series of extraordinary innovations in later decades, notably the discovery of a form of encryption that underpins all our internet transactions. For obvious reasons, little can be said about more recent breakthroughs, but GCHQ's reputation within government for devising and delivering innovative and complex technology and solving difficult problems at scale is unique. While government tends to be a byword for IT inefficiency, underperformance and overspend, GCHQ, increasingly in collaboration with industry, generally bucks that trend.

This book does not hold up Bletchley Park and GCHQ as perfect examples of creativity, innovation and high performance. There were – and are – numerous mistakes, dead-ends and failures. But it tries to explore the conditions for their remarkable success; what is it about the people of these secret organisations – their selection, development, organisation and culture – that enables innovation to flourish? What have we learnt from the problems posed by sophisticated adversaries in this dark world, from the creators of Enigma to modern terrorists and cyber-criminals? Above all, what are the lessons that might be transferrable to the leadership and management of organisations trying to adapt and innovate at speed in this century? How do we create and sustain the magic we need to meet escalating complexity?

Bletchley was described at the time as a 'rudderless ship', an 'asylum' and 'almost complete chaos', yet by D-Day it had become the world's largest intelligence factory, giving Allied commanders better insight into the enemy than at any time in military history. GCHQ has been described in equally unflattering terms, and yet has become a natural centre of gravity for skills in a cyber-enabled world where technology dominates.

Identifying the strand that runs through this history is not an exact science. It is less about mapping the DNA of Bletchley Park and more like identifying the ingredients of a digital sourdough starter: a messy, blended fermentation that constantly changes, that is never entirely within the baker's control, but nonetheless produces something consistently surprising. The answers will not be in the form of yet another prescriptive management system, but rather a deliberately eclectic series of observations and provocations, from which any reader can take what he or she likes. This book takes a very roughly chronological approach from the 19th century onwards, but disappears into numerous rabbit holes along the way. There is some method in this chaotic and unsystematic approach, which I hope will become clearer by the book's end, as the puzzle fits together.

1

Through the Looking Glass

To enter GCHQ's headquarters in Cheltenham is a disorientating experience for most visitors. Leaving mobiles, laptops and smart watches behind removes your attachment to the usual barrage of digital information and connection with the outside world, and to a sense of time. Stepping through automatic doors into a glass security booth and then out into the cathedral-like marble and granite atrium is the first of a series of surprises. The sense of disorientation grows as you experience the circularity of the building, with its never-ending internal 'street'. Surprisingly open-plan spaces are punctuated by a mixture of the hyper-secret and the humdrum, a Costa or Starbucks franchise juxtaposed with advanced research into a new technology. Step beyond this hadron collider of busy activity and there is a haven of quiet greenery at the centre of the Doughnut, where you could be in a cloistered garden bar the hum of the vast data centre below.

A newspaper editor, on his way out of this building after a series of briefings and increasingly agitated by his separation from his electronic devices, told me that he felt he had fallen down the rabbit hole into 'an *Alice in Wonderland* world'. It was an unintentionally perceptive reference. A hundred years earlier, Britain's early codebreakers had marked the conclusion of their efforts in the First World War by writing a parody of Lewis Carroll's story, based on the idea that Alice had fallen down a chute into a basket full of encrypted messages. The long poem poked fun at the many curious characters she encountered working in 'Room 40' of the Admiralty and their puzzling world of secret problem-solving (so sensitive that the parody itself remained secret for decades).

Nor is it a coincidence that Charles Dodgson, the man behind the pen name Lewis Carroll, is still an inspiration at GCHQ. A mathematician and teacher (among other things, he taught one of the early female codebreakers at school), he had an extraordinarily diverse range of interests, from novel democratic electoral systems to new methods of encrypting messages. A problem-solver and amateur engineer, Dodgson enjoyed inventing all sorts of practical devices, including a book holder for bed-bound patients and an ingenious aid to writing in the dark (he often woke with an idea he wanted to record, so he kept this device under his pillow to avoid having to get out of bed and light an oil lamp). His inventions were the solutions to particular practical problems, the products of an engineering mindset. But Dodgson was also a great puzzle-setter, perhaps the greatest; he invented a game that was a precursor to *Scrabble* and many of the word games he came up with are still popular. It is this thread of puzzling that runs through the labyrinth of GCHQ's work and organisation, stretching back to the 19th-century birth of big data and forward to the technological advances of intelligent machines that are still being imagined, designed and put to work.

Pulling on this thread leads you beyond puzzles as a diverting pastime on your daily commute, to the setting of deliberately difficult challenges: secret codes that are designed only to be cracked by those with the right key, and, of course, in a race against time – cracking codes that are used by bad people to hide bad things for bad reasons.

Continue pulling on the thread and you begin to see that the puzzles GCHQ is trying to figure out are much broader than codes. Intelligence work is like solving a complex jigsaw from tiny fragments, perhaps trying to prevent an imminent terrorist attack knowing only part of a phone number, or a misheard name, or a vague description. This is a jigsaw puzzle where most of the pieces are missing, the size of the picture is unknown, and even the pieces in front of you may be broken or from another puzzle altogether. Or it may be that there is simply too much information – too many pieces from too many jigsaws – and the challenge is to see a pattern and sift out the noise before it is too late.

A final pull on the puzzle thread and it frays in some fascinating directions. The discovery of mathematical patterns in nature all around us consumed Alan Turing, GC&CS and GCHQ's* most famous employee, and many others during the last century. Faced with an ocean of apparently indecipherable Enigma-enciphered gobbledegook transmitted across the radio waves, a mindset of pattern discovery and the estimation of probability was central to the success of Bletchley Park in the Second World

* GC&CS stands for the Government Code and Cypher School, formed at the end of the First World War and renamed GCHQ in June 1946.

War and to the programming of the first computers built there. In this unremarkable part of Buckinghamshire, science, mathematics, engineering and the humanities sustained national survival.

Puzzles and conspiracies

Back in Cheltenham, following the puzzle trail led me in two directions, one theoretical and one practical.

The more I delved into the world of puzzles, the more I appreciated the overlap between recreational puzzling, GCHQ's core mission of countering threats and the deeper human compulsion to do puzzles. We are after all the only species that sets puzzles, although other animals have been observed solving them. The desire to see patterns, to make sense of the baffling, has been a recurring theme throughout human history and seems buried deep within. Ultimately, it is about making sense of the world around us.

As I began to look at the rise of online conspiracy theories and see the motivation of some of the terrorist groups I had dealt with, it was clear that this psychological programming towards puzzle-solving and pattern recognition also has a dark side. Our innate desire to see order and coherence in everything, to make sense of the puzzles of life, is the flip side of a deep fear of randomness, difference and inexplicable coincidences. The more shocking and traumatic an event, it seems, the more we need an intelligible and satisfying explanation. This fear leads very often to some remarkable distortions. Many of the most bizarre conspiracy theories, from Covid vaccines to 9/11, seem to come from people with deep grievances and profound fears, which compel them to arrange facts in such a way as to bring their own version of order to chaos. Of course, conspiracy theories arise and spread for many reasons. But seeing the event as a puzzle, to which an intricate and sophisticated arrangement of facts is the solution, satisfies conspiracy theorists, for whom complexity is attractive. They need to see an intricate pattern, whether or not there is one.

What codebreakers and puzzlers of the GCHQ variety have to offer is more than simply winning the codebreaking or puzzle-solving race. They can also teach us about challenging these false patterns. Code making and breaking are essentially scientific exercises that have to be based on an underlying truth if they are to be useful; there must be a clear method for decoding a message, one that is testable and obtains the same result each time, and the result must be the message originally intended by the author. This is an obvious practical requirement. Imagine the opposite: if, for example, encrypted instructions for firing the missiles of a nuclear submarine could be read in a wide variety of ways, subject to the emotional disposition of the reader, the world would be a good deal riskier.

To put it another way, the long-running television puzzle show *Only Connect* (favourite viewing for some in Cheltenham) works because there is only one correct answer from a limited pool of possibilities. If, instead, everything can be connected to everything else at the whim of the puzzler, it ceases to be a meaningful exercise or much fun as a game. There has to be a truth somewhere, which is one of the things the serious side of puzzling can teach us.

Practical puzzlers

Yet beyond contemplating the origins of the human instinct to set and solve puzzles, there were some very practical questions to ask in Cheltenham. What kind of people, with what skills, are best suited to this work? What mixture of abilities and personalities and brains work best? How are they best recruited, organised and managed, and what sort of workplace will they need to flourish? Is the mindset of curiosity, problem-solving and innovation teachable? How can it be applied outside the secret world to the great challenges of the future?

In looking for organisational answers, there were many lessons from the history of GCHQ, some of which I have pursued in the stories below. One conclusion was perhaps surprising: most of the country and most of the readers of this book have an aptitude for this work. This came home to me in my second year at GCHQ, when I decided to mimic in my Christmas card Bletchley Park's use of the *Daily Telegraph* crossword puzzle as a recruitment tool.

In January 1942 GC&CS worked with the *Telegraph* to mount a competition for avid crossword puzzlers, the challenge being to complete the whole thing in under twelve minutes. Some of those who succeeded were then privately approached by GC&CS staff with a view to taking up this specialist war work. Of course, this was only one short-lived recruitment experiment and there were many other routes to Bletchley, but it gave an insight into both the skills required for the work and the diverse pool of possible candidates.

The purpose of the 2015 Christmas card was not to recruit new staff to GCHQ, although that was not ruled out. Apart from paying homage to Bletchley Park codebreakers, the main objective was to increase public understanding of the organisation, to remind people of the importance of having problem-solvers working in national security, and to raise money for mental health charities and the NSPCC.

I knew that there was a very active group of recreational puzzlers in Cheltenham, primarily among the mathematicians and cryptanalysts, and so I asked them to come up with a puzzle that could be printed on our Christmas card and made available to the public through the GCHQ website. It needed to be in several stages and difficult enough not to be solved quickly, but it ought to have something accessible to everyone. The final stage should be so difficult that only a small number of winners would arrive at the solution. Ideally, and to avoid internet crowd-sourcing of answers too quickly, which would spoil the fun for the majority, it should not be completely clear whether the winner had got the final answer right. This being government, there would be no significant prize, beyond the kudos of cracking it and a cheap paperweight.

To my amazement, and without any obvious promotion, over 600,000 people tried the puzzle online, with 10,000 getting through to the final stage and six coming very close to the solution. The competition was so popular that we decided to collect the many puzzles created over the years by GCHQ staff for their own amusement and to publish a puzzle book; it was – and remains – a bestseller, and, along with its successor, has raised over £500,000 for mental health charities. Its popularity illustrated what we know from opinion surveys, which suggest that three-quarters of the population regularly try puzzles of some sort. From Sudoku to Wordle, jigsaws to treasure hunts, mystery novels to TV detective shows, this is a national hobby in most countries. Even Queen Elizabeth II, when she visited our new National Cyber Security Centre, revealed that she spent a lot of time on complex visual puzzles, especially very large jigsaws.

So a journey through GCHQ's history is a journey shared by many people who might never imagine being in the secret world. It is about mysterious numbers, words, pictures, objects and even sounds, and how to make sense of them. It is about conspiracies and mysteries, and the thin line between puzzles and paranoia. It is about the serious business of puzzles and codes, and how to solve and unravel them, but above all about the kinds of people – often unexpected people – who do this. Not surprisingly, the best way to explore this strange world, and what it can occasionally teach us, is to disappear down a series of rabbit holes.

2

Round the Bend

I used to keep two paperweights on my desk at GCHQ. The first was a small chunk of the internet. It was an undersea cable about 10cm thick, and surprisingly rigid and heavy. Most of that was steel armour to protect it against the hazards of the ocean floor, and insulation to keep water and electric currents apart. At its centre was a group of hair-like optical fibres, each capable of carrying massive volumes of information over vast distances. I kept it to wave at visitors as an illustration of the reality of bulk data – the data being carried along these tiny fibre-optic cables – and as a reminder to me that the internet is profoundly physical, made up of tubes and cables and racks of computers in cooled data centres all over the world. It has a geography and a history, and is built and operated by people. There is nothing 'cloud'-like about it.

In October 2023 the Swedish government revealed that undersea internet cables between Estonia and Sweden had been damaged; they could not rule out Russian involvement, and indeed the former Russian president Dmitry Medvedev had already suggested that Western cables were fair game for attack. The same year, outlying islands of Taiwan appeared to have been cut off from the flow of information. Governments in the West have begun to worry publicly about the fragility of the global internet on which every aspect of their societies and economies depends. How many cables coming into the UK would need to be cut before the City of London ceased to operate?

Reliance on a transnational spaghetti of some 1.5 million kilometres of mostly privately owned cables criss-crossing the ocean floor suddenly looks like a problem. There are dozens of breaks to these cables every year, usually caused by fishing boats or mechanical failure, but deliberate sabotage seems new.

Cables themselves are not new, of course, any more than information warfare is a recent phenomenon. The first transatlantic cable came ashore at Porthcurno in Cornwall (now the site of an extraordinary museum) in the 1880s; its modern equivalents emerged close to GCHQ's facility at Bude in the same county, a coincidence that became the source of endless conspiracy theories.

At the outbreak of the First World War Norwegian engineers inspected their end of the Anglo-Norwegian undersea telegraph cable, which had ceased to function; they found a fake electrical device had been added. It was there to confuse them about where the break in the cable was, and inside was a handwritten note that read:

No more Reuter war-lies on this line! Kindest regards from a 'Hun' and a 'Sea Pirate'.

By 1915 the 'cable wars' were well underway. In fact, within a few hours of the declaration of war the previous year the British cable-laying ship CS *Alert* set sail from Dover, and after midnight began cutting German telegraph cables running through the English Channel to France, Spain, the Azores and on to the United States. Dredging up the right cables with grappling hooks in the dark and in bad weather was not straightforward, but by the end of the operation Germany's telegraph communications with the Americas and Africa had been severed. After the transatlantic cables, *Alert* cut another six cables linking Berlin with London.

Over the years that followed, both sides cut numerous cables, including those to neutral powers such as Norway. From the Pacific to the Indian Ocean, the North Sea to the South Atlantic, cable cutting became more widespread and more sophisticated. German raiding parties attacked cable-landing and relay stations as far away as the Cocos Islands south of Indonesia and Fanning Island in the Pacific, even developing remarkably advanced techniques for U-boats to cut cables in deeper water.

The result of this was not simply to disrupt the enemy's ability to communicate, although disruption was an added benefit. The Admiralty came to realise that it was forcing German communications onto the widespread but relatively new technology of radio, which the British could intercept more easily. The choice for Germany was between this technology and using indirect wired telegraph connections through London, which the British censor would be able to intercept. By cutting cables, the communications space could be 'shaped' and made manageable.

These two themes – owning the communications infrastructure and protecting the information travelling across it – which we tend to think are unique problems for the 21st century, are in fact old concerns. From finding the right people, skills and understanding for the data economy, to the broad issues of privacy, surveillance, business models and geopolitical rivalry, we have been there before.

History offers no easy solutions for today's technological challenges, but it can dispense some wisdom. The disruption to trading nations and companies as communications became electronic in the 19th century was as jarring in its volume and speed as anything we see today. For governments, a new type of arms race had begun. Keeping information secret at scale – and breaking down that secrecy – demanded new people, new approaches, new ways of thinking. The modern intelligence world grew out of this struggle and helped to shape the future. To the extent that it succeeded, it found a new kind of employee who blended scientific expertise in the still newish technology of electricity with problem-solving or engineering skills. That sort of person could not be produced by the traditional educational engines of Empire. To see how it happened we need to travel back to the Middle East in the mid-19th century – and go round the bend.

'Telegraph Island'

The tiny island of Jazirat al Maqlab in the Persian Gulf can still be reached by traditional dhow, the sail-driven cargo workhorse of the Indian Ocean. Adventurous tourists go there to snorkel, dolphin-watch or admire the stunning scenery of Musandam, a vast peninsula of rock where the Arabian and Eurasian tectonic plates meet, but not many stay long in this unsheltered spot in high summer, when temperatures routinely reach 40°C. Some steps and a scattering of low-level ruins are the only clues that the island was once inhabited.

The *Illustrated London News* of 8 July 1865 gave a rather breathless spin on the delights of this as a workplace, mentioning that

> *servants essential to European life in such a climate have also their quarters, so that there is no monotony at the station from lack of human life. Boats are provided for exercise and amusement, and a regular supply of English periodicals and newspapers ... the time passes away very quickly at Mussendom.*

In a hint that it might be more public-relations exercise than balanced assessment, the author added that 'two hulks are fitted up for the staff to live on board whenever a change from the island is preferred'. In reality, life in this blisteringly hot space the size of two football pitches, situated in a tiny airless inlet between high cliffs, was oppressive. The young Scottish engineer

Patrick Stewart wrote that 'in a purely sanitary point of view' it would be better to move the facility because 'the heat ... the high encircling rocks and limited view to seaward must have a depressing effect upon Europeans, especially during the hot season'.

In recent years etymologists have suggested that the nautical phrase 'going round the bend' refers to this island, since going around the corner of the peninsula in the Gulf to spend time on the island meant a trip to insanity. The station closed within three years.

But visits were necessary because, as 'Telegraph Island', this rock was part of a near-global network of undersea and over-land cables that revolutionised international communications in less than two decades – a breakthrough on a par with the creation of the internet and a direct antecedent to it. It was on this network that the first electronic signals were propelled across continents at high speeds. An early telegram on the Indo-European cable might take just less than half an hour to get from Bombay to London, although operators could cope with no more than twenty words per minute. But set against a sea-journey time of months, this was a world-changing invention.

Technology has a history, albeit one that is rarely prominent in the school curriculum, and to understand intelligence gathering in the 20th and 21st centuries this context is crucial. Spying itself is as ancient as humanity – it features in written sources from at least the Old Testament onwards – and not surprisingly revolves around attempts to hide and discover sensitive information. But while human nature may not have changed much, the transmission, availability, scale and speed of communications have altered almost beyond imagination in the past 150 years.

On the eve of the looming global wars of the 20th century, human spies and their governments found themselves working in a world where science and mathematics were changing fundamentally the way information was created, processed and communicated. The methods by which the secret world operated were about to become exponentially technical, and success would go to those most adept at the new disciplines. This would not occur without a struggle during the last century of the millennium, as the comfortable practices of the old order began to be disrupted by new kinds of people.

Increasingly, the means by which communication took place moved out of the direct hands of governments, starting with the private sector consortium that launched the first transatlantic telegraph cable to the current multi-company world of undersea cables carrying the global internet, the providers of encrypted services and the leaders in artificial intelligence. But what is less well understood is that crucial developments in the technology that enabled this new digital world, and a key part of what now helps people to stay private within it – whether for good or malign purposes – stem from a history in which the secret and military worlds played a crucial role.

The great age of 19th-century science saw the introduction of electronic communication over long distances by telegraph; radio followed, freeing people from cables; and after the Second World War satellites and, above all, fibre-optic technology enabled the expansion of communications on a huge scale. The birth of the internet, with its origins in the US defence world, 'democratised' what had previously been the preserve of governments and corporations; hand-held computing power – the direct descendants of the enormous computers created by intelligence staff at Bletchley Park – facilitated by advances in miniaturising transistors, brought a new wave of mobility.

Alongside their obvious benefits, all of these advances have presented extraordinary challenges in the 21st century for those in the intelligence community whose job it is to look for bad things: to find, follow and tackle the minority who want to use modern technology for harm. Within a decade everyone had moved on to the same technologies, so the intelligence puzzle grew from finding a needle in a haystack to finding a piece of hay in the haystack. The relatively static lines of communication along which single messages were transmitted suddenly changed; the internet worked by messages being broken down into 'packets', then fired around the world through multiple routes before reassembly at the receiver's end. A parallel from an earlier age would be the difference between intercepting and opening a letter, and reconstructing that same letter from numberless paper particles spread across the globe.

At each stage of this rapid advance, the secret and military worlds were not merely bystanders hoping to adapt to a new context; at their best – and somewhat haphazardly – they were creating, innovating and helping to drive the technology itself. They therefore inevitably invented some of the problems they subsequently had to face, including the widespread use of high-grade encryption. The benefits, however, outweighed the risks.

In 1914, almost by accident, the secret world created a whole new discipline of 'signals intelligence' to cope with the challenges posed by electronic communication, mechanisation and the increasing sway of mathematics over natural language. An important role for GCHQ, its Bletchley predecessors and American counterparts, was – and still is – to understand the future trends in communications and place bets on which technologies would emerge dominant. As the 'signals' changed and moved, so those trying to find and read them had to reinvent their craft. The understanding of how Morse code is transmitted electronically along a 19th-century copper cable is radically different from the propulsion of 'packets' of internet data as light along a fibre-optic cable, even if some of the concepts at play are the same.

Signals intelligence is therefore by definition dynamic and future-facing. It tries to predict what communications technology might look like in twenty years' time, and how it may be used or misused by those who wish to cause harm. As we shall see, when the intelligence community got this right in the last century, amazing things happened and advances were made that themselves benefited wider societal progress.

But intelligence work is also precarious, an area of government that seems forever on the point of being left behind, going out of business, becoming obsolete. Keeping ahead of the curve is essential to its continued ascendancy. How this has been achieved, by maintaining the culture and people necessary for success, is the subject of what follows.

Science under fire

A good place to start is back with the young Patrick Stewart in the Persian Gulf. Although a long way from being an intelligence officer – a role that scarcely existed even in formal military structures at the time – he was a prototype of the kind of public servant that would be needed in GCHQ and the future struggles of the secret world.

In 1848, in his mid-teens, Stewart became a cadet at Addiscombe, the East India Company's training college for its own military forces. His friend Frederic Goldsmid tells us a great deal about his schooling, including the fact that the institution was almost out of control, with 'frequent disturbances' from unruly cadets who regularly ran amok in nearby Croydon.

While its sister institution, the East India Company College, trained imperial clerks and administrators in calmer Hertfordshire, Addiscombe focused on mathematics and science, the curriculum at this subsidised college being determinedly practical. It trained problem-solvers and engineers, and its demographic was much closer to the school leavers, apprentices and engineers who became the backbone of GCHQ in the 20th century.

Stewart himself excelled at 'mathematics and fortification', and he soaked up the excitement of the many advances in engineering and mechanisation all around him. When given a sneak preview of the Great Exhibition of 1851 he was dazzled by the 'splendid pieces of machinery'. He toured civil engineering sites across England on a school trip, stopping off at Clifton to be shown around the construction of the new suspension bridge by Isambard Kingdom Brunel, its chief engineer and architect.

This period was every bit as exciting as the internet era, both seemingly times of infinite possibilities. By the age of nineteen Stewart was overseeing the building of bridges in India, as well as being effectively in charge of developing the first telegraph lines across this vast country, extending them some 1,700 miles from Calcutta to Lahore and Agra to Indore. But Stewart's contribution to the history of electronic communication was not only to perfect the surveying and construction of cables in one part of the world. He helped the military and civil authorities understand the value of electronic communications and the perils of not having control of them.

His arrival in India coincided with widespread unrest and eventually, in 1857, a military rebellion against the rule of the East India Company. His role during the Indian Mutiny, as the rebellion was called, was both practical – building fortifications – and tactically critical to military commanders in making communication possible through telegraph cables.

The great *Times* correspondent W. H. Russell witnessed Stewart at work and described his role in support of Sir Colin Campbell, whom Lord Palmerston had sent out to India as commander-in-chief to tackle the uprising. Stewart 'put the end of the telegraph wire into Sir Colin's hand wherever he went', and for the first time electronic communications kept pace with an advancing army, giving Campbell almost real-time information on the progress of enemy forces, as well as the possibility of controlling his own over long distances. The advantages to a commander of this kind of telegraphy were obvious; much of the ultimate success in raising the siege of Lucknow and suppressing the revolt was attributed to the superior communications of the British.

But this also put telegraph operators in harm's way, with Stewart himself enjoying some narrow escapes as he worked within rifle range of the enemy. The active use of electronic telegraphs for military communications demanded a new kind of employee – someone from 'the scientific branch of the army', as Stewart was described, who could do science while under fire. Russell's florid prose described Stewart and his men being attacked by cavalry, and although their electric batteries and transport were destroyed by gunfire, electronic pulses were nonetheless transmitted across the copper wire:

But still they work on, creep over arid plains, across watercourses, span rivers and pierce jungles, till one after another the rude poles raise aloft their slender burdens and the quick needle vibrates with its silent tongue amid the thunder of the artillery.

By the 1870s the 'C' Telegraph Troop of the Royal Engineers had been formed, operating in the Anglo-Zulu War of 1879 and soon developing into the 'Telegraph Battalion' that operated across the Empire.

Military engineers professionalised this service to the point that eventually a self-standing Royal Corps of Signals became justified and necessary. It is no coincidence that it was Winston Churchill, as secretary of state for war, who signed them into existence in 1920. Cables, telegraphy and radio had played a key part in the First World War for the navy and the army, with Churchill being among the first to spot their strategic importance and understand the opportunities they presented.

The birth of GC&CS in 1919 was deeply intertwined with military communication, and for the following hundred and more years the organisation has supported military operations. A significant proportion of GCHQ's staff has always been supplied by the armed forces, something that suddenly becomes obvious in Cheltenham when they all wear uniform. The relationship works in both directions: GCHQ supporting military objectives, and the military enabling GCHQ's collection of intelligence in difficult places and conditions.

In one of the larger rooms of the Doughnut in Cheltenham there is piece of traditional military art – an oil painting of members of the Royal Signals climbing up a communications tower to fix a microwave communications dish during the capture of Basra in southern Iraq in 2003 while under fire from both air and ground. This, along with many more clandestine military operations, stands in a direct line with the actions of Patrick Stewart and his contemporaries.

But the events in Lucknow in which Stewart participated held an importance well beyond the Indian Mutiny. The sense of national scandal and the political recrimination in London that followed the uprising led the British government to conclude that rule of the subcontinent could no longer be outsourced to the East India Company, with more direct control from Westminster requiring greater and faster communication in both directions. Those who had already been arguing for an

Indo-European telegraph cable were emboldened and, more importantly, funded.

Patrick Stewart was put in charge of this project, and in particular of the Persian Gulf cable in 1862. As well as energetically surveying the route and procuring the cable, he mastered what became a recurring theme in the 20th century – the geopolitics of electronic communications. At various moments over the next fifty years other routes were sought, as the passage through Persia was seen as too risky and there was deep suspicion that Tsarist Russia would read the cables. Handing over the nation's communications to other states and overdependence on particular regions are not simply modern preoccupations. The British government explicitly wanted all parts of the Empire to be connected by British-owned cables precisely so that it did not have to rely on cables it did not control. This concern has come back to haunt modern governments as they contemplate their dependency on cables laid and owned by both tech companies and 'difficult' states.

Stewart did not live to see the completion of all his tasks, which included the extension of a cable towards China through Singapore, as he died in Constantinople in 1865 at the age of thirty-three. Although a brilliant engineer, he was accident prone – as Goldsmid delicately put it, 'He had acquired a great local reputation as the most unfortunate of beings in respect of hurts and tumbles.' He sustained lasting injuries from a wounded tigress who carried him around in her jaws for a while, despite Stewart pretending to be dead so that she would lose interest. The following year he had a spectacular riding accident that caused some sort of brain haemorrhage, followed by a near-fatal racket-ball injury that severed an artery. This was in addition to the fevers that traditionally went with a posting to India (one of Stewart's last scientific tasks was to lead a commission to investigate the prevalence of cholera in India, a disease from which he had suffered himself).

Stewart was one of a new generation of technical public servants, a 'scientist' who straddled the military, administrative and political worlds. His deep knowledge of electrical engineering was only part of his success in making the rapid completion of the Indo-European link seem so effortless. He was an excellent project manager and assembled a remarkable team from the close-knit world of the telegraph cable constructors: manufacturers like Siemens and the Gutta Percha Company.

In contrast to the acrimonious history of the transatlantic cable, where Samuel Morse and a succession of eminent experts resigned or were sacked, Stewart's team maintained a strong sense of purpose. His contemporaries describe a self-effacing man, uninterested in the glory of his achievement, who handled foreign leaders and his own staff alike with great tact. He created an organisational culture that prefigured the amalgam of science, industry, military and government service we will see at Bletchley and in parallel developments in the United States.

This young group – most of them in their twenties – worked in the new technology of electricity, an invention as novel to them as the internet was to many of us until a few decades ago. Building on the rise of electrical engineering in the London of Faraday and Maxwell, Stewart and his colleagues helped construct a global communications network of enormous strategic significance. They were the intellectual ancestors of those who would devise the internet and operate over it a hundred years later, and they foreshadowed the practical workforce of the new secret world.

It was to this historical parallel that we turned when the government sought to enact new legislation after the Snowden disclosures of 2013. David Cameron as prime minister wanted the maximum possible transparency about what powers the intelligence agencies – particularly GCHQ – possessed to intercept communications, and what limitations, safeguards and oversight were in place, especially in the area of bulk internet data.

We were consequently asked to educate the political classes and the public in a new way about what we did, even if there were obvious limits to what we could say about the techniques used or the information gathered. A succession of senior politicians and newspaper editors trooped over to Cheltenham, and our briefings often started with a semi-serious slide that our chief technologist called 'The Victorian Internet' – a 1912 map of telegraph cables around the world, alongside a modern sub-ocean internet cable map. It made the very obvious point that bulk data and interception had a history (although it occasionally led to genuine confusion, one visitor from the House of Lords genuinely wondering whether they had attributed to Silicon Valley something that was actually Victorian).

The map also demonstrated some wider truths. The cables on the sea floor mirrored global trade routes, with the main early connections across the Atlantic and from Europe to India. Most critically for the development of signals intelligence, they converged on London, the centre of the British Empire. The repeater stations also explained some of the odder colonial preoccupations, not least St Helena and Ascension Island in the South Atlantic, both of which were staging posts for the cable laid in 1899 to improve communications between London and Cape Town during the Second Boer War. On Ascension Island, the cable came ashore at the appropriately named 'Comfortless Cove', a remote spot once used to quarantine infectious sailors when they stopped for resupply. I once visited it en route to the Falklands and found it well named.

The map showed the advantage Britain enjoyed by being at the centre of such a complex communications web. Its pre-eminence in the construction and operation of cable-laying ships, along with control of Malaysia, where the rubber-like gutta-percha insulating material was grown, gave London a head start, added to which was the fact that many of the world's cables had to be routed through England. Since by far the largest and most important of these were transatlantic, it also illustrated the centrality of the Anglo-American relationship in the global communications net.

At the beginning of this chapter I mentioned the cable paper-weight on my desk. My other paperweight was more bespoke, produced to mark the end of Operation Herrick, the UK's military campaign in Afghanistan against Al-Qaeda and the Taliban from 2002 to 2014. In the tradition of 'trench art', it was a small piece of crushed data centre equipment, mounted in plastic, commemorating the efforts of GCHQ staff both on the ground with the military and back in the West Country. It also pointed to the dominance of the communications technology used by the Western allies. When GCHQ pulled out with the armed forces after 2014 it publicised the destruction of its vast data infrastructure in Afghanistan, a massive logistical exercise in crushing equipment – and then crushing the crusher for good measure. Owning and shaping the technical space was a lesson learnt a hundred years earlier; in the 21st century that ownership has passed predominantly to private industry, and with it many of the challenges and responsibilities experienced by governments in the preceding century.

With the advent of artificial intelligence, which staff at GC&CS did so much to create, and the imminent arrival of quantum computers, governments can no longer shape, still less own the environment. But there remain options for staying in the game and focusing on those who will continue to misuse technology.

3

To Utopia and Beyond

Back in 2014 tech companies were very much in public favour. Google, Facebook and Apple were destinations of choice for young engineers and many others, and the iPhone was only seven years old at the time. The companies involved in this new era of personal computing and communications were hugely fashionable and almost beyond criticism. Government, by contrast, was in the dock after the Snowden leaks, accused of all sorts of nefarious data gathering.

On my second day in the job I had an article published in the *Financial Times*, which, slightly naively, I thought was a mild statement of the obvious. I simply pointed to our current struggle against ISIS and suggested that they were a new kind of terrorist group. Where their more antiquated Al-Qaeda predecessors had used the internet as a traditional means of communicating, ISIS had 'embraced the web as a noisy channel in which to promote itself, intimidate people, and radicalise new recruits'. They were the first terrorist group to master disinformation online.

My article provoked an avalanche of reaction, occasionally constructive but usually not. Most of it missed the point entirely, claiming that this was a plot by intelligence agencies to invade privacy. By contrast, what was clear from the perspective of the secret world was that the bad things going on were not just our problem, they were everyone's. We simply happened to be the ones tasked with tackling them.

Things look different now – the tech giants have fallen from grace, accused of collecting and monetising more data of and about us than any government ever could. Scandals have mounted up, with these companies being accused of everything from profiting from teenage suicide to the monopolistic crushing of competitors. There is a widespread acceptance that 'surveillance capitalism' is unhealthy, even if the alternative internet model is unclear. And many people instinctively feel that there has been some vast social media experiment on the younger generation, the results of which, good and bad, will only be understood with the passage of time – and probably when it is too late.

But the fundamental question has not changed. If the tech companies are not going to fix the bad things going on across the internet and governments cannot, then who will? In lengthy engagements I undertook with well-meaning people about electoral disinformation, the misuse of encryption and poor cyber-security standards, I found that the private sector and civil society were sympathetic but ultimately abdicated responsibility. I suspect there are two reasons for this. One reflects the culture of Silicon Valley and other hothouses of cutting-edge tech. If you are wealthy and successful and live in Palo Alto, you are less likely to lead the kind of life where you suffer the most extreme negative consequences of the internet, although you and your family are hardly immune from the pressures of the online world. Additionally, those involved in technological progress have tended to be optimists and utopians. They have built their vision on a model that either assumes consistently good human behaviour or ignores the dark side altogether.

The second reason is commercial. Where bad things are noticed at all, corporate pressures are such that it makes more financial sense to look the other way. It is of course true that an over-emphasis on safety at an early stage stops innovation in its tracks; as Mark Zuckerberg remarked, the advent of manned flight would never have happened on that basis.

But safety did become important when flying was scaled up and made available to the public. The challenge that the tech companies – and indeed all of us – have failed to meet is that the spread of the internet and the dissemination of social media have happened at such a pace and on such a scale that safety concerns have been left far behind. Commercial drivers dominate and steamroll over any concerns.

The secret world, however, is quite different. It spends all its time looking at those who will abuse technology and do harmful things. It tries to anticipate how they will do it and then tries to stop them. It cannot afford to be utopian.

An important role of agencies like GCHQ has been to challenge the natural and unthinking idealism that accompanies technological advances. The inventors and engineers of the commercial world are drawn instantly to sunlit heights and dazzling possibilities, not to mention all the imagined financial rewards. Of course, there are plenty of experts in GCHQ who are similarly excited by the advances of technology, particularly the ones who have contributed to that progress. But those in the secret world whose job it is to deal with the darker side of human behaviour – to prevent bad people doing bad things – have another perspective. They are the voice – sometimes the lone voice – in the corner at the over-hyped product launch asking quietly whether there might actually be some downsides as well as benefits. Freed from commercial and political pressures, and with the credibility that comes from understanding and sometimes even creating these technologies, they possess a unique questioning function.

In GCHQ's case, this function derives from a long history of trying to protect information and to break open things that others are trying to hide. Following this back down the rabbit hole takes us to those early years of global connectivity, a time of huge opportunity and excitement, yet also one that illustrated the need for technologies to keep information safe and expose its misuse by others.

'Let us be one'

As we have already seen, in the second half of the 19th century the arrival of a global net of telegraphy generated wonderful visions of a united world. Rudyard Kipling wrote of these 'shell-burred cables' in 1893, pointing to the collapse of time and the potential for worldwide harmony:

They have wakened the timeless Things; they have killed
  their father Time
Joining hands in the gloom, a league from the last of the
  sun.
Hush! Men talk to-day o'er the waste of the ultimate slime,
And a new Word runs between: whispering, 'Let us be one!'

The first political message dispatched across the Atlantic, from Queen Victoria to President James Buchanan, sent on 16 August 1868, was suitably bland. It spoke of 'common interest and reciprocal esteem'; its ninety-eight words took sixteen hours to send and were exceptionally hard for the Morse code operators to decipher. By contrast, Buchanan's reply was much more effusive, worthy of the most optimistic cheerleaders of the internet; he hoped that the undersea cable would 'prove to be a bond of perpetual peace and friendship between kindred nations', spreading civilisation, liberty and law throughout the world. It was '[an achievement] more glorious, because far more useful to mankind, than was ever won by conqueror on the field of battle'.

Think of Mark Zuckerberg's confident promise that Facebook would build stronger communities and that 'by giving people the power to share, we're making the world more transparent', and you can see the same undiluted idealism about technology.

But we have already seen that the cable wars at the start of the First World War dented 19th-century hopes. And just as knowledge and data transfer were crucial in the military and government spheres, they were equally so in the commercial world. Paul Reuter realised that profit came from transmitting knowledge faster than anyone else, and his telegraph and news company succeeded by getting information across the Atlantic first. Before the transatlantic cable was completed, a combination of telegraph, ship and strategically positioned pigeon services ensured that Reuter could get news from a transatlantic steamer ashore in Ireland and onwards to London before rival steamers reached Southampton. Reuter's agency anticipated the automated traders of our own era, for whom high speeds and low latency signify the difference between success and failure. Algorithmic automated trading systems, issuing multiple buy and sell orders, have replaced humans in many areas; in these trades, an advantage of a millisecond over a competitor's trade may be worth tens of millions. How fast light travels along the cable therefore matters, as well as who might interrupt it or even cut the cable.

In a pre-echo of our own times, Reuter was accused of misusing the monopoly he enjoyed by delaying the news of Abraham Lincoln's assassination in 1865 to manipulate stock prices. It was the exploitation of this kind of commercial advantage that inevitably undermined the fantasy of global unity – Reuter had guaranteed an annual fee for his use of the cable, just as modern tech companies rent or own internet cables.

President Buchanan's optimism about the benefits to civilisation of rapid communication was not wholly misplaced, but it was certainly naive. Nation states – as well as companies – quickly sensed the opportunity to be both competitive and aggressive in new, electronic ways, and the possibilities for manipulating information began to proliferate.

The technology utopians were so excited by the possibility of a global instant communications network that they started speculating about a universal language that could be used to send messages across it. This was not a new dream. The great 17th-century English scientist John Wilkins (in whose lodgings at Wadham College, Oxford, I am writing this) explored the idea in 1668, without any prospect of a practical means of communicating with the other side of the world.

Presented with this new electronic global network, a German priest, Johann Schleyer, set out to invent a global language, motivated by a parishioner's difficulties getting the US Postal Service to understand his writing. Schleyer wanted this language to be 'capable of expressing thought with the greatest clearness and accuracy', the learning of which should be 'as easy as possible to the greatest number of human beings'. He called it 'Volapük' ('worldspeak') and it was, very briefly, an extraordinary success.

Spreading quickly across Europe in the 1880s, the language had as many as a million active speakers; there were Volapük grammars in multiple languages and nearly three hundred Volapük clubs sprung up around the world. Then the vision died almost as quickly; splits emerged between purists led by Schleyer, and pragmatists who wanted the most useful possible language for commerce and science.

This Volapük rabbit hole is worth mentioning not just because it illustrates the spirit of excited utopianism, but because of the role of one of its leading enthusiasts in setting the ground rules for encryption and codebreaking, which were to carry through to Bletchley and beyond. Whether we all spoke with one language or many, here was the problem: if massive volumes of information could be fired around the world, how could any of it be safe or private?

The principles of secrecy

Auguste Kerckhoffs, a Dutch teacher of modern languages, was to have a profound impact on how secrets were kept and transmitted across this new technology. Kerckhoffs was a linguist and polymath, at various times a professor of German and French, but whose work and interests covered art, science and theology. His focus on linguistics and the science of language explains his enthusiasm for constructing a new, practical universal language, but it also led him to investigate the secrecy of communications and to set out some core principles that still guide practitioners in the secret world.

While teaching in Paris, Kerckhoffs looked at the codes and ciphers being taught in French military schools for securing messages. He was not impressed, concluding that the failure to invest serious effort in codebreaking meant that no one was really testing the codes currently in use. If they had done so, even an amateur cryptanalyst would break them quickly.

He set about writing one of the key guides to the problem of secret communication, *La Cryptographie militaire*, published in 1883. In the two papers that comprised this work he drew on his broad knowledge of language and a great deal of research about cryptology to frame six practical principles for secret communication in the era of global telegraphy. Kerckhoffs realised, based on his reading of the progress of technology, that all future means of encrypting information needed to be able to cope with the volume and speed of communications, and therefore needed to keep changing. He also saw that the only way to ensure good encryption was to become very good at decryption – to test to destruction the chosen methods for securing information. This has been a guiding approach for GCHQ.

Kerckhoffs's six principles still underpin modern approaches to encryption. His first stated that the system chosen should be unbreakable in practice, even if not in theory, correctly concluding that a tactical and immediate military message might still be sensitive long after it was sent. This principle therefore amounted to saying that encryption needed to be good enough to withstand attack for whatever period the material needed to stay secret; theoretical perfection was less important. Principles three to six showed his very practical understanding of human beings operating encryption: keys should be memorable and easy to change; encoded messages needed to be transmissible by Morse code; the system ought to be portable; and it must be simple to use without extensive training or expertise.

All these envisaged a machine that could take the strain off the user, and be easily changed. But Kerckhoffs' key insight came in his second principle, which stated that the design of the system should not itself require secrecy. In other words, in a world where communications were proliferating and many people in many geographies would need to know how to secure information, we should assume that the system or method would become known to the enemy. The answer had to be that the specific keys used, and regularly changed, would be the guarantee of security and secrecy. It need not matter if the basic code system or machine were captured, because replacing them all would be impossible. A system needed to be designed so that it could withstand that compromise. That was true of Enigma machines but is equally true of the computer I am writing on. I should assume that a cyber attacker understands how my machine works, how I am protected and how my encryption works in principle, without being able to break it.

In looking at codes and ciphers from a fresh perspective, Kerckhoffs thereby came up with the basic principles that would lead the way to mechanised systems like Enigma. Although he also proposed some particular methods of encryption, his real contribution was conceptual – understanding from first principles what needed to be achieved.

Looking back for a moment to the momentous history of technology in the 19th century, its military and secret applications before the First World War, and at the origins of our contemporary problems, we can see two themes that underpin everything that follows, both the challenges and their solutions.

The first is volume and speed. By the outbreak of war in 1914, we had already entered a high-speed world, driven by the march of science and awash with information flowing around the globe. Governments were faced then with exactly the same challenge as presented by the cyber age. How could they guarantee the confidentiality of information so that it was read only by the intended recipient? How could they preserve its integrity and stop it being altered? How could they protect the availability of this electronic information, on which so much economic, military and governmental activity depended? These questions – of confidentiality, integrity and accessibility of data – remain the core framework of the modern approach to cyber security.

For governments, ensuring that information would remain secret became a major preoccupation, starting an arms race of encryption and decryption that continues to the present day. While the use of codes and ciphers – the hiding or obscuring of information – had existed since at least classical antiquity, it was now suddenly propelled from the era of individuals and dilettantes into the world of physics, mathematics and industrial mechanisation.

4

Fragments in the Sand

I once spent some time during a particularly long Whitehall meeting, appropriately along the corridor from the original First World War codebreakers' base, Room 40 in the Old Admiralty Building, reading articles on the neurological benefits of a nice warm bath. Studies on rats in the last twenty years apparently suggest that immersion in warm water causes serotonin-releasing neurons to fire in the brain, potentially altering mood, concentration and intellectual performance. Taking a bath, apparently, is good for the brain.

In 1914 there had in fact been a bath installed in a neighbouring office to where I was sitting at the request of one of the great figures of signals intelligence, Alfred Dillwyn 'Dilly' Knox. It was long gone when I was in Whitehall, but Knox's story and his academic background as a professional papyrologist, piecing together fragments of ancient manuscripts found scattered in the desert sands of Egypt, tell us a lot about the origins of GCHQ. Knox was one of a band of complete amateurs in a codebreaking profession that scarcely existed before the First World War and which had to be created from scratch.

The emergence of GC&CS at the war's end is the story of solving a large-scale but poorly understood problem. As intercepted Morse code radio messages piled up, what kind of skills were needed to decode them? What sort of people had these skills? And once decoded, how could this information be best used? The solutions stumbled upon by these pioneers at GC&CS, a group who were interested in the problem as a puzzle or game, are instructive. They had an almost blank canvas and they began to create something revolutionary.

It might be assumed that the Admiralty's speed in cutting German telegraph cables on 4 August 1914, soon after the outbreak of the First World War, pointed to a well-prepared and sophisticated British plan for reading and using their communications, but that was not the case. Initial collection of intercepted German radio traffic was haphazard and partial, collected by amateur radio enthusiasts, companies like Marconi, and a growing network of military sites, including the Wireless Telegraphy Station at Scarborough, which later became GC&CS's first station and is probably the oldest still-functioning signals intelligence collection site in the world. It was set up in 1912 as a Royal Navy communications station, but it also monitored radio transmissions from the newly created High Seas Fleet of the German Imperial Navy, a key part of Admiral Alfred von Tirpitz's plan to challenge the British navy – then thinly spread around the Empire – by concentrating forces in the North Sea and harrying the east coast of England.

The Scarborough station sums up the adaptability and development of the signals intelligence mission over the 20th century. From monitoring the German navy in the First World War, it subsequently performed a key role as part of the network of 'Y' stations intercepting naval and Luftwaffe radio transmissions for Bletchley Park, playing a significant role in the tracking and sinking of the *Bismarck* in 1941. From 1945 it was repurposed to collect the radio traffic of the Soviet armed forces, especially the navy. Scarborough was a key source of information on Soviet intercontinental ballistic missiles and contributed to US intelligence during the Cuban Missile Crisis in 1962. Radio interception still matters, but the focus has shifted to the cutting edge of signals, in particular training for the cyber world, making this northern seaside town a symbol of the rapid evolution of technology-based intelligence, always adapting and trying to keep ahead of the enemy.

But at the outbreak of the First World War little thought had been given by government to the decryption of all these radio messages being somewhat randomly scooped up in Scarborough and elsewhere. The recruitment of a number of remarkable people and the gradual organisation of the effort led to some spectacular successes. While not used as fully as they might have been by a navy that either did not understand them or could not use them in a timely manner, the decoded signals nonetheless proved a concept. They pointed to the kind of people and systems that would be necessary at Bletchley Park, and laid the groundwork for the formation of what became GCHQ. These sometimes chaotic efforts found in Winston Churchill, as First Sea Lord, a politician who was not only excited – sometimes over-excited – by the detail of the messages themselves, but who understood the game-changing possibilities of signals intelligence.

Churchill had always been fascinated by secret intelligence and took a number of personal risks in the Second Boer War attempting to gather it himself. Throughout his working life he had a voracious appetite for detailed and 'raw' reports, unfiltered by others. Looking back in 1923 to the naval battles of the First World War, he contrasts the silence of Whitehall with action on the high seas:

> *only the clock ticks and quiet men enter with quick steps, laying pencilled strips of paper before other men, equally silent, who draw lines and scribble calculations ... telegram succeeds telegram at a few minutes' interval as they are picked up and decoded, often in the wrong sequence, frequently of dubious import; and out of these a picture always flickering and changing, rises in the mind, and imagination strikes out around it at every stage flashes of hope or fear.*

This excited fascination could be burdensome to those producing the material and irritating to his commanders as he second-guessed them. But it had a number of advantages for those seeking the priority, funding and personnel they needed. It also meant that Churchill understood better than most the importance of protecting the sources of intelligence, something that would become a necessary obsession at Bletchley and in the modern era.

Amateur codebreakers

The early and growing successes in codebreaking came about by the inspired choice of some unusual staff. In August 1914, with unreadable coded German messages piling up on his desk, the head of naval intelligence turned to a physicist and engineer for help. Sir Alfred Ewing was already well known and at the peak of his career. Born in the era of Patrick Stewart, he had plenty of international experience of the practical application of science. As a student he had accompanied the great William Thomson – later ennobled as Lord Kelvin – on a cable-laying expedition to Brazil, and he became one of the first foreigners to serve as a professor at Tokyo Imperial University, teaching mechanical engineering during Japan's modernisation programme in the 1870s. In this earthquake-prone part of the world he was instrumental in creating one of the first seismographs, returning later to Scotland, where, as professor of engineering at Dundee, he combined research on magnetism with practical overhauls of the city's appalling sewage system.

Ewing knew little about codebreaking, but his role as the newly created director of naval education seemed to allow him plenty of time and freedom. He had been dabbling unsuccessfully in the design of cipher machines, and on being set the task of decrypting some early messages he spent time in the British Library studying the history of cryptology. Most importantly, he brought together several friends and colleagues, including Alastair Denniston, a teacher at Osborne Royal Navy College, who was to become a key figure in the development of technical intelligence and the first director of GC&CS.

Ewing was given freedom to recruit and organise as he wished, not least because he was too senior and distinguished to be under anyone else's authority. Following the outbreak of war in July 1914 he drew in a few volunteers, mostly from the faculties of the naval colleges during what was their summer vacation. None of these were codebreakers – no such profession existed – and their only qualification was some knowledge of German. They made very little progress in the first weeks, as the small team worked surreptitiously in Ewing's office and intercepts continued to multiply. They then had an extraordinary piece of luck – of a type that has characterised many subsequent breakthroughs in signals intelligence. Churchill provides the context in his history of the First World War, describing what happened in August 1914 after the German light cruiser SMS *Magdeburg* ran aground and was then struck by Russian naval vessels off the Estonian mainland:

> *The body of a drowned German under-officer was picked up by the Russians a few hours later, and clasped in his bosom by arms rigid with death, were the cipher and signal books of the German Navy, and the minutely squared maps of the North Sea and Heligoland Bight.*

The Russians offered the code books to the Admiralty, who sent a ship to collect them. 'Late on an October afternoon,' wrote Churchill, 'Prince Louis [of Battenberg] and I received from the hands of our loyal allies these sea-stained priceless documents.'

'Priceless' was no mere hyperbole. Possession of the books was a phenomenal breakthrough, but there was further work to be done before messages began to be decrypted in November that year. Churchill recalls that they were 'mostly of a routine character', detailing ships' positions and movements, but by assembling them the Admiralty could at least in principle see the layout of parts of the German High Seas Fleet.

By this point the group of four temporary summer helpers had grown and they required new accommodation, moving to Room 40 in the Old Admiralty Building on Whitehall. If anywhere is the geographical ancestor of Bletchley Park and GCHQ, this is it, which in some regards is a pity as these days it is a nondescript and not easily identified room, following many decades of office remodelling. But from this obscure and innocent-sounding location the group started to enjoy significant successes; these were noted by Churchill, who would regularly ring up to congratulate the team.

Missed opportunities during the subsequent years of the war, in particular confusion surrounding the use of Room 40's intelligence at the Battle of Jutland in 1916, and the increase in volume of material as the flow from the network of radio intercept stations grew, led to expansion and reorganisation. Over the years that followed, Room 40, working with its War Office equivalent MI1b, broke most of the German and other codes given to it. Particularly important was the work against early U-boats, a precursor of Bletchley Park's critical work to protect convoys during the Battle of the Atlantic. Over time, Room 40 and MI1b decrypted diplomatic traffic from almost every combatant government, including the United States. The volume was large; those involved estimated that 15,000 secret messages had been decoded by the war's end, but many more intercepts had been received, perhaps as many as 2,000 daily. Handling this was a purely human rather than mechanical task, for the last time in the history of British and American wartime codebreaking.

The criteria for recruitment to Room 40, which quickly spilled out of the room but retained its eponymous title, was to have good German, be intelligent and trustworthy. This was a rarer combination than might have been expected, and there was no clear method for identifying candidates for jobs that themselves had few well-defined parameters. Add to this the absence of young men, most of whom had already volunteered for active service on the Western Front, and recruiters had to make do with the unpredictable and often physically unfit flotsam and jetsam of society. In fact, one of the great underlying stories of GC&CS is how it performed extraordinary feats with a collection of people no one else wanted or considered important.

Ewing and his extraordinary successor, Admiral William 'Blinker' Hall, recruited a bizarre mixture of people. Ewing turned to academia and brought in a number of Fellows of King's College, Cambridge, including Frank Adcock, Frank Birch and Dilly Knox. These were neither scientists nor mathematicians, despite Ewing's own background, but he clearly believed that a papyrologist like Knox would be useful.

The study of papyrus fragments was a relatively new discipline in classical studies, following the excavation of large volumes of papyrus pieces from the sands of Oxyrhynchus in the Nile delta. These revealed the lost works of Greek and Roman authors, including further 'gospels' depicting the life of Jesus. But reconstructing and deciphering tiny sections of writing was like a painstaking, complex jigsaw puzzle. Knox specialised in the fragments of the Greek author Herodas from the 3rd century BC. Significantly, Knox had no great love for Herodas' bawdy poetry. Mavis Lever, one of Knox's team at Bletchley, recorded that he treated the poet 'as a familiar foe, and [his poetry] provided a difficult game in which nearly all the rules were missing, one which Dilly intended to win and arrive at the correct text'.

Given Knox's later importance with respect to Enigma and other tasks at Bletchley, it is worth remembering why this unusual civilian was so well suited to the fuzzily demarcated and little-understood skill of codebreaking. Working on the papyri was much more complicated than a visual jigsaw challenge of piecing together some bits of paper from the sand, although Knox complained that hours staring at the fragments in the British Museum destroyed his eyesight. And, of course, there was no template for what a poem should look like; the lid of the jigsaw box was long gone. The fragments of manuscript rescued from the sands had not been written down by the author Herodas himself, but copied out by a scribe three or four centuries after his death. Often these scribes were not particularly competent, or they were bored or careless or did not understand what they were copying. In the case of Herodas' poetry, the scribe that Knox studied was not only prone to all the usual errors, 'but worst of all he suffered from a schoolboy knowledge of Greek and, where he followed the sense roughly, made uncon-sciously stupid alterations'. To complicate matters further, Herodas himself enjoyed switching between two different ancient Greek dialects for poetic effect, deliberately surprising the reader.

Deciphering the fragments therefore involved understanding both the poet and the idiosyncrasies of the scribe – the human errors that would be so important at Bletchley. If the scribe was the adversary, this meant understanding how he thought and behaved. Knox could also use external factors – notably the rhythmic structure of the poetry – to guess where the scribe had gone wrong. More importantly, when he found a reliable passage, this became a 'crib' or cheat on which he could base his decipherment of the rest. After the First World War he adopted a similar approach to tackling Hungarian diplomatic codes: in the words of his biographer and Bletchley colleague Mavis Lever, 'He did not trouble to learn the language but treated the whole thing as an abstract problem.'

Knox's translation of Herodas was still in use when I studied classics sixty years later. In fact, I got to know many of the Room 40 and Bletchley codebreakers through their publications in ancient history and classical languages – rediscovering them as codebreakers in the secret world was a strange homecoming.

Knox would prove Ewing's original instinct right, becoming a pivotal figure at Bletchley Park and leading an almost all-female team. Knox excelled in his first excursions in cryptanalysis in Room 40 (strictly speaking it was Room 53, because that is where he had his bath installed). Room 53 became his office, shared later with his wife-to-be Olive Roddam, and he allegedly cracked a German naval flag cipher while soaking in the bath. This was not a whim or affectation – twenty-five years later he spent long hours in the bath at the cottage in Bletchley Park where he and his team did their work (so long and in such deep contemplation that on one occasion an anxious colleague broke the door down, fearing that he had drowned).

Knox found that taking baths aided the 'perception of analogies', without revealing why. The recent research into the benefits of immersion in warm water cited earlier may explain the neurological effects, although Knox's bath was probably not warm for long either in the Admiralty or at Bletchley. But the calming effect of water and isolation may nonetheless have been important for a man who was passionate about the challenge before him and found it hard to conceal his frustration when solutions were not self-evident.

It is tempting to dismiss the Room 40 occupants as a random group of eccentrics and dilettantes. While it is true that there was an array of Late Victorian or Edwardian talent from famous families, serious work was unquestionably being done. Hall recruited a large number of women as clerks – perhaps for the first time in intelligence work – and organised the Room 40 operation into separate sections, eventually coordinated with MI1b at the War Office. A mixture of lawyers, bankers, stockbrokers, actors and publishers joined the academics, and by the end of the war seventy-four men and thirty-three women were working in Room 40.

In MI1b there had been a similar expansion. As an example of how unprepared the country was for this new kind of electronic intelligence, in 1914 the War Office had expected its solitary librarian, Francis Huddleston, to be a one-man code-breaking service for the entire British army. Of the female staff in MI1b we know relatively little, except that there were linguists and translators as well as typists, and they were paid less than their male counterparts. One of them, Clara Spurling, specialised in Scandinavian languages. She had been educated at Oxford High School for Girls at a time when the Rev. Charles Dodgson (aka Lewis Carroll) was a tutor in mathematics. Clara was reported to be the only applicant to MI1b ever to achieve 100 per cent in the near-impossible entrance test set by its head, Major Malcolm Hay. She fared much better than her male counterparts, some of whose less modest, self-promoting applications survive.

Mirroring Room 40, a small group of civilians and military officers was drafted in to help decipher the growing volume of telegrams captured through the censorship system, which enabled the government to hold and inspect all communications travelling across telegraph cables. In the years that followed, MI1b branched out into the interception of radio communications, with outposts overseas, and it enjoyed notable successes against German Zeppelin airships. At the height of the First World War, MI1b was reading encrypted diplomatic messages ranging from Argentina and Japan to Persia and the Vatican.

The Zimmermann Telegram

Among the many ground-breaking moments for intelligence gathering in the war, the decryption of a single telegram by Room 40 stood out as illustrating the geopolitical impact that reading the enemy's communications could have. It proved the concept of signals intelligence at a strategic as well as tactical level, and the original hand-written translation is still one of GCHQ's most prized possessions.

By the end of 1916 stalemate on the Western Front and the blockade of shipping into German ports caused a radical change of policy in Berlin. Germany determined that unrestricted submarine warfare, including against American ships, was the only way to stop the resupply of Britain and shift the balance of the war. As part of this the German Foreign Ministry sought to contain the damage that might be caused if such action brought the United States into the war.

Faced with the consequences of the cable wars – the lines of communication severed by both sides – the German Foreign Ministry had found new ways to get communications to a still neutral United States, first by having the Swedish Foreign Ministry send them as their own telegrams and later by persuading President Woodrow Wilson that German use of American diplomatic channels would support his efforts for peace. Wilson agreed, with the condition that messages should not be encrypted ('in clear', and therefore readable).

Using this route, and with some spectacular cheek, Arthur Zimmermann, the German foreign minister, sent an encrypted telegram to his ambassador in Mexico on 16 January 1917. In it he instructed the envoy to suggest a deal to the Mexican president, the substance of which was politically explosive, especially in offering up parts of the United States as payment.

We intend to begin on 1 February unrestricted submarine warfare. We shall endeavour in spite of this to keep the US neutral. In the event of this not succeeding we make Mexico a proposal of alliance on the following basis: make war together, make peace together.

Generous financial support and an undertaking on our part that Mexico is to reconquer the lost territory in Texas, New Mexico and Arizona.

Since all telegrams from neutral countries in Europe were routed through London, this was immediately brought to Room 40 and by the following morning Nigel de Grey, a young man who was to play a major role in the running of Bletchley Park, had made a first attempt at reading it. Recruited for his proficiency in French and German, de Grey had worked for the publisher William Heinemann and would later return to civilian life to run Medici greetings cards.

Hall immediately understood the explosive political impact this telegram might have in tilting the balance of debate in Washington towards the US entering the war on the Allied side. But it presented a dilemma that became a recurring theme of the use of intelligence over the next century and up to the present day. Revealing that German code had been broken would obviously lead to better encipherment and possibly the end of readable traffic. Added to this was the delicate problem that Room 40 was reading US diplomatic messages, the route used to transmit the telegram from Zimmermann.

Hall's solution was to distance the decryption from Britain and from signals intelligence. By obtaining a copy of the telegram in Mexico, where it had been forwarded from Washington, and spreading word that a spy had been involved, Hall was able to present the telegram to contacts in the US embassy in London a month later, describing it as something that had been bought in Mexico and decrypted there by a clever British diplomat. When the story became front-page news in America it seemed entirely unrelated to Britain, offering no hint that German codes had been involved or broken.

What Hall achieved through personal inventiveness was to be turned into a vast disinformation campaign to protect the precious secret sources of Bletchley Park. Hundreds of thousands of messages at Bletchley had to be 'sanitised' to ensure that, if they were seen by the enemy, it would not be obvious that they had come from intercepted and decrypted communications. Elaborate cover stories involving spies were then disseminated to make the German military machine suspect betrayal by humans. The mantra of secrecy, which Churchill had enjoined on Room 40, was to become an extraordinary blanket of silence spread over the operations at Bletchley until the 1970s.

It would be an over-simplification to say that the Zimmermann Telegram brought the United States into the war, but it certainly hastened the end of American isolationism in this period. It was important ammunition for those who wanted to argue against even-handed neutrality in Washington, all the more so because

Zimmermann publicly admitted that the telegram was genuine
– today he would probably have questioned its authenticity and
blamed a 'fake news' conspiracy against him.

Hall himself was not an aristocrat or an academic, nor was
he a cryptanalyst. Neither was Ewing, who freely admitted that
he could not begin to compete with the team he had helped
create. But Hall knew how to use intelligence for the nation's –
and his own – good and how to conceal its source. He came
from a naval family and had enlisted at the age of fourteen.
Nicknamed 'Blinker' for his ever-present facial tick, he possessed
an extraordinarily forceful energy. The American ambassador in
London wrote to President Wilson to describe the experience of
meeting Hall: '[He] can look through you and see the very
muscular movements of your immortal soul while he is talking
to you!' Hall had established a close relationship with the
ambassador but was also reading his secret telegrams – the
'special relationship' was still some way off.

Whatever his personal qualities, particularly his tendency to
promote himself, Hall understood the role of the leader of an
intelligence enterprise. He recruited the right people and delegated to them. Giving his Second World War successors advice
that was not always heeded, he said:

A Director of Intelligence who attempts to keep himself
informed about every detail of the work being done cannot
hope to succeed: but if he so arranges his organisation that he
knows at once to which of the organisation he must go for the
information he requires, then he may expect good results.
Such a system has the inestimable advantage of bringing out

the best of everyone working under it, for the head will not suggest every move: he will welcome and indeed insist on ideas from his staff.

This would have been a radical approach to leadership even within the military sphere, but given that Hall meant his ragbag of civilians, it was all the more revolutionary. He grasped two points, however, that proved central to the success of Bletchley Park and which all subsequent heads of GCHQ have had to address – how to enable 'ideas from staff' who understand more than the director does, and how to keep abreast of the sheer volume of information generated by such a vast and disparate enterprise.

As the First World War drew to a close, most of Room 40's occupants prepared to return to civilian life. They celebrated by performing a lengthy satire on *Alice's Adventures in Wonderland* – *Alice in ID25* (the new name for Room 40, although rarely used) – written by Dilly Knox and Frank Birch, the latter being one of the more extrovert characters in the team. Birch had served in the navy before joining the cryptanalysts in 1916, then went on to have a successful career in the theatre and in some well-known films of the 1930s before leading the German naval section at Bletchley, finally becoming head of GCHQ's historical section (as well as resuming his film career) until his death in 1956.

The poems in the satire were written by Dilly Knox, and the whole skit revolved around the theme of the cryptanalysts' surreal world – Alice comes across an encrypted intercept message, falls down a pneumatic tube and finds herself in the cage of texts waiting to be decoded by the strange assortment of creatures assembled in Room 40.

The pantomime, which was not made public until the 1970s, says a lot about the era, the personalities and some of the key moments of the previous few years, including the Zimmermann Telegram. It reveals a collegiate and determinedly amateur world of intelligence production, stuck somewhere between Victorian Britain and the 20th century. But as war was being industrialised and professionalised, so too was the business of intelligence, and the years that followed provided an opportunity to test what would be necessary in the future.

As an experiment, Room 40 had demonstrated that intelligence gathering and codebreaking could be done at speed. It had proved beyond doubt the value of tactical material – for example, the direction-finding radio interceptions that enabled the Admiralty to triangulate the position of enemy ships. Additionally, in the case of the Zimmermann Telegram, it had shown the strategic and political importance of intercepting non-military traffic.

It also, however, exposed some weaknesses. Room 40 and MI1b produced large volumes of decrypted messages, and the use of human systems – some 200 people spread across the two organisations by the end of the war – was undoubtedly efficient. But they were micro-organisations that had not yet made the leap from impressive cottage industry to industrial enterprise. Mechanisation was on the way.

5

The American Way

Many years ago there was an internal staff Christmas pantomime put on by the three intelligence agencies – MI5, MI6 and
GCHQ. As we have seen, such productions have a pedigree
going back to 1918 and the *Alice in ID25* satire written by the
early codebreakers of Room 40. In contrast with the raffish
metropolitan MI6 officers or Smiley-esque counter-intelligence
investigators, the GCHQ employees in sketches were carica-
tured as geeks (or in the old days, 'boffins'). But they were also
seen as the introverted and socially awkward cousins from the
sticks – not part of the urbane world of gentlemen's clubs and
Whitehall intrigue – and were usually depicted staring down at
their feet (Bletchley veterans made similar observations about
the cryptanalysts walking around the Park, totally absorbed in
their own thoughts).

These caricatures have never much bothered the targets of the
satire – after all, in an age in which geeks, nerds and techies are
ruling the world, they can afford themselves a certain quiet
satisfaction. But it may come as a surprise to many to realise
that the creator of the James Bond mythology, Ian Fleming,
drew heavily on his experience of the world of codebreaking

and technical intelligence. In particular, he seems to have modelled much of Bond's behaviour, from alcohol and poker to his attitudes to women, on one of the most entertaining scoundrels of the American signals intelligence world.

Operation Ruthless

Fleming's role in naval intelligence gave him some insight into what had been achieved in Room 40. It is even possible that Bond's 007 codename is a reference to the Zimmermann Telegram, the greatest coup of the early codebreakers, which had an internal codename of 0075. Fleming was one of a small group who had some knowledge of what was happening at Bletchley Park in the Second World War. At the worst point of the Battle for the Atlantic, when German U-boats were sinking huge numbers of merchant ships, pressure to help Alan Turing and his colleagues with shortcuts to breaking the traffic encoded by Enigma intensified. The desperate need for daily settings and tables from naval Enigma machines led to equally desperate ideas. Fleming devised a plan – codenamed Operation Ruthless – to support the work being done at Bletchley in 1940 that is worthy of one of his novels.

He proposed to his boss that he would get hold of an airworthy German bomber that he and four others would crew, dressed in Luftwaffe uniforms. They would need to speak German and be 'tough, bachelor, able to swim'. The aircraft would join the tail end of a German bombing raid as it returned from London towards France, and then deliberately crash into the Channel, after radioing for help from a German naval rescue boat.

Fleming and his crew would then take over the rescue boat, with its Enigma machine and codebook, and head for home. If things went wrong, as seemed highly likely, their cover story would be that they were have-a-go heroes who had taken the plane without permission and wanted to take the fight to the enemy. The fact that Fleming's Admiralty bosses thought Operation Ruthless worth contemplating showed how desperate the need was. In the end, the operation was shelved after no suitable rescue boats were identified, much to the disappointment of Alan Turing, who had taken a childlike delight in the idea of the escapade.

Fleming's enduring inspiration for the personal character of Bond probably came from his exposure to the turbulent career of another codebreaker, Herbert O. Yardley. In the course of a rollercoaster life, Yardley wrote a sensational, bestselling 'whistle-blower' book, worked for at least three governments and ended up writing a still-famous guide to playing poker. He also spectacularly exposed the ethical issues raised by invading the privacy of other people's communications and, a century before Edward Snowden, enraged opinion by publishing a detailed account of his work as a spy. He was both honoured and threatened with prosecution by his own government, and despite everything is buried in Arlington National Cemetery.

Yardley first learnt about telegrams from his father, a railway worker and telegraph operator in rural Indiana. He dropped out of university and joined the railroad himself as a telegrapher, a job that required Morse code, good literacy and a knowledge of railway terminology. Having passed a civil service examination, he moved to the State Department as a junior clerk in December 1912, enrolled in a post that required little thought; he simply had to register incoming telegrams from US embassies around the world and put them in the appropriate boxes for senior officials to read.

Bored stiff during the long night shifts, Yardley decided to see whether he could understand any of the encrypted messages. He discovered he could read several of them without too much effort and with a little research into books on cryptology he found in Washington libraries. He was particularly alarmed and excited by one discovery:

> One night, business being quiet, I heard the cable office in New York tell the White House telegraph operator … that he had five hundred code words from Colonel House to the President. As the telegram flashed over the wire, I made a copy. This would be good material to work on, for surely the President and his trusted agent would be using a difficult code.

Within a few hours Yardley had exposed some very poor presidential security, and went on to write a lengthy report on how American diplomatic messages might be better protected. As the United States entered the First World War he was keen to apply his new understanding to enemy codes, so he approached the head of military intelligence, Major Ralph van Deman. Just as the War Office in London had assumed before July 1914 that a single departmental librarian could handle all the codebreaking, so there had been equally scant preparation in Washington. Van Deman had outsourced work by sending encrypted messages to a small team at Riverbank Laboratories, Illinois, of whom we will hear much more in Chapter 15; but he saw Yardley's report and request for a posting, and commissioned him into the Army Signal Corps, creating a one-man unit – Military Intelligence-8 (MI-8).

Yardley set about building a cryptanalytical group, hiring a number of academics from the University of Chicago, who would have felt quite at home in Room 40. John Manly and Edith Rickert were early recruits; they met in MI-8 and went on to spend twenty years together at Chicago after the war on a monumental project to analyse the many manuscripts and printed editions of Chaucer. Rickert's book on the application of a scientific approach to textual analysis, *New Methods for the Study of Literature*, published in 1922, reads like a code-breaker's handbook. Her method searches for patterns and uses statistical analysis; rhythm in poetry, for example, is 'a succession of sound-groups created in accordance with some unifying psychological principle'.

Manly and Rickert were responsible for MI-8's most high-profile achievement, the successful conviction of a spy responsible for one of the worst acts of terrorism in US history, and the largest before 9/11. Lothar Witzke was a saboteur, part of a German intelligence ring that was responsible for blowing up naval munitions in San Francisco and along the western seaboard in 1917. But the explosion at Black Tom Island in New York harbour on 30 July 1916 was on a completely different scale. The island housed a depot for arms and ammunition for export to Europe; 45,000 kilograms of TNT and nearly a million kilograms of ammunition were ignited, sending shockwaves across New York. Windows throughout lower Manhattan and as far away as St Patrick's Cathedral on Fifth Avenue were shattered, and the blast was felt in Maryland and Connecticut. Closer to the island, the torch of the Statue of Liberty was seriously damaged – and never reopened to visitors – and Ellis Island immigration centre had to be rapidly evacuated.

A major investigation was launched, but American domestic intelligence capabilities in 1916 were, like their interception of communications, somewhat rudimentary. In fact, the incident led almost directly to the establishment of federal investigative agencies such as the FBI and the introduction of legislation to criminalise espionage.

Following a tip-off, Witzke was arrested on the Mexico–Arizona border two years later posing as Pablo Waberski, a Russian American. An encrypted text was found sewn into his coat, which was eventually forwarded to MI-8. After some months, Manly and Rickert, under Yardley's direction, cracked what was an extremely sophisticated cipher. The message read:

The bearer of this is a subject of the [German] Empire who travels as a Russian under the name of Pablo Waberski. He is a German secret agent. Please furnish him on request protection and assistance; also advance him on demand 1,000 pesos of Mexican gold and send his coded telegrams to this Embassy as official consular despatches.

The author of the message was the same hapless envoy of the German Empire in Mexico, Heinrich von Eckardt, who had already been the recipient of the Zimmermann Telegram that was decoded by Room 40. Manly was a witness in the trial that sentenced Witzke to death, although his sentence was commuted after the war and he was deported to Germany in 1923.

While basking in the glow of this achievement for his unit, Yardley's 1918 visit to London to exchange information was less successful. He was warmly received in the War Office, but Admiral 'Blinker' Hall would not allow him to visit Room 40. Hall seems to have regarded Yardley as indiscreet, which he certainly was, but it is as likely that Room 40's regular decryption of American diplomatic telegrams was the bigger problem.

'Gentlemen do not read each other's mail'

As the war ended, Yardley's MI-8, like its London counterparts Room 40 and MI1b, had a great deal to celebrate. They had proved the concept and value of this new area of signals intelligence in building the complete picture of the enemy to which Churchill had aspired.

The creation of the Government Code and Cypher School (GC&CS) was mirrored by the desire in Washington to continue its own cryptological unit. Although funded by the military, the Cipher Bureau moved into an anonymous office block in New York, just off Fifth Avenue. New York was chosen largely because the commercial cable and radio companies were located there. This was the start of Yardley's problems and set off a dispute over governmental access to data that has continued to run through intelligence gathering to the present day.

Outside the peculiar conditions that occur in wartime, there was simply no legal basis for cable companies to hand over telegrams to Yardley, although his personal charm and their sense of patriotism meant that company bosses found ways of giving him what he wanted, and in time cables were being copied and transported to the Bureau in large volumes every day. In order of priority, Yardley's targets were Japan, Britain, Mexico, Germany and Russia. After a huge effort, Yardley's team cracked Japanese codes in time for the Washington Naval Conference that started in November 1921. Their successes earned Yardley the Distinguished Service Medal and, according to Yardley's papers in the NSA archives, a Christmas bonus for his staff.

But by the end of the decade the Cipher Bureau was gone and Yardley was out of work. There were a number of reasons for this, each of which points forward to some of the themes of intelligence work. For all the successes of 1921, the Bureau had not delivered game-changing material in subsequent years. It had failed to spot the trends towards the mechanisation of encryption that led to machines like Engima. Yardley had not read the advance of technology in the same way as his friend and rival William Friedman, who worked in the War Department in Washington, or Alastair Denniston in London. Nor had he ensured a new flow of talent into the Bureau with the right skills for the mechanical age. He had also neglected the relationship with the military, which wished to develop its own codebreaking capabilities.

But what actually finished Yardley was an ethical decision by Herbert Stimson, appointed as secretary of state by the newly elected President Hoover in 1929. In contrast to British and other European foreign ministers, who showed no equivocation about reading intercepted material, Stimson demurred, his idealism genuinely held and shared widely across the political divide in Washington. His famous phrase 'Gentlemen do not read each other's mail,' in response to Yardley's proud recitation of intercepted messages of foreign counterparts, was not about manners but based on deeper conviction. Intercepted intelligence material might be necessary in wartime – and in fact Stimson as secretary for war was a heavy user of it from 1940 – but it was 'highly unethical' for the State Department to be doing this in peacetime. To Yardley's bemusement, his intention to break into Vatican communications seemed to be particularly egregious.

If this illustrated the deep divide between US and European attitudes to privacy and what was acceptable in espionage, Yardley's decision to write a tell-all book united all governments in disapproval. Jobless and struggling at the start of the Great Depression, Yardley turned to writing out of necessity, although he also claimed he was drawing attention to the poor state of American security. *The American Black Chamber*, published in 1931, was an international bestseller, probably the only one in the history of cryptology. This name for the US cipher bureau was a nod towards the 'black chamber' of Sir Francis Walsingham, Elizabeth I's spymaster (and for many years 'The Walsingham' was the nickname of the 'pub' that opens for a few hours once a week in GCHQ), and a reminder that interception of messages was not new. But it set out in great detail the mass copying of telegrams in breach of federal law, revealing how many ciphers of foreign governments had been broken and outraging opinion in Japan in particular.

Despite being full of exaggerations, inventions and self-promotion, the book was great storytelling. Yardley became a celebrity, but opinion was deeply divided on what he had done. For a long time the intelligence community regarded him as reckless at best, as well as responsible for improvements in Japanese cipher security in the 1930s.

Over the years, calmer assessments prevailed regarding the damage caused. Frank Rowlett, the man more than anyone responsible for breaking the Japanese 'Purple' cipher machine before Pearl Harbor, took the view that Yardley's book 'helped us more than it hurt us'. Japan's premature transition to 'Purple' actually gave the US time to work on it during the 1930s and left them in a stronger position at the outbreak of war.

In London there were unsuccessful attempts to suppress the book; the intelligence establishment was already used to fire-fighting revelations about Room 40 by Admiral Hall, Sir Alfred Ewing and even Churchill in previous years. In this there was a strange reverse symmetry; while the British were much more relaxed about 'reading gentlemen's mail', they were less keen on a public debate about it. The opposite was true – and to some extent still is – in America.

The rest of Yardley's life was an endless pursuit of income and the hobbies he enjoyed, primarily cryptology, travel and poker. He worked for Chiang Kai-shek in China and briefly for the Canadians. There he set up the 'Examination Unit' in 1941, Canada's first codebreaking bureau and the forerunner of the

Communications Security Establishment (CSE), Canada's modern equivalent of GCHQ. Canada played an increasingly important role in the effort against German, Japanese, Italian and, of course, Vichy French codes, and had recruited academics along the lines of its counterparts in Britain and America. Yardley's arrival in Ottawa created more alarm in London than Washington, and Denniston eventually intervened to have him replaced, probably because he feared for the safety of the Enigma secret, with some justice given Yardley's track record. Instead, he sent Oliver Strachey out to Canada. He was a veteran of Room 40 and a talented cryptanalyst but, at sixty-seven, some decades older than anyone else involved.

Yardley was still frozen out of Washington, where the law had been changed to prevent repeats of the *Black Chamber* experience. His final book, *The Education of a Poker Player*, was a bigger bestseller than his first and remains widely available. It combines practical advice with extraordinary anecdotes based on the characters he had met in shady bars around the world.

A number of famous amateur and professional poker players, among them David Mamet, Victoria Coren Mitchell and Al Alvarez (who wrote an introduction to one edition), have cited Yardley's work as their route into the game. It is in small part an instruction manual, based on the thousands of hands the author had played over the decades. But it is more a collection of what Ian Fleming regarded as 'the finest gambling stories I had ever read ... the book had zest, blood, sex and a tough dry humour reminiscent of Raymond Chandler'. It is easy to see in Yardley's accounts of his adventures in the dingy back rooms and bars of Chongqing, where he had been hired to run a code-breaking bureau for the Chinese nationalist government of Chiang Kai-shek against the advancing communists, the proto-type for Bond's best and worst characteristics. The gambling stories are full of eccentrics and baddies worthy of a Bond film.

Experienced poker players now regard the book as a good read but an outdated view of the game. Yet Yardley saw in poker, as Bond does, the essence of intelligence work. In Alvarez's words, Yardley shows that the game is 'not about luck but calculation, memory, patience, skill in reckoning the odds and percentages and above all, observation: the ability to recognise and interpret the small fidgets and quirks, the hesitations, the voice's faint changes in timbre which indicate tension or confidence'.

This ability to blend calculation, memory and probability, along with human observation, explains the brilliance of the early amateur codebreakers. Notwithstanding his personal flaws and Hollywood fantasies, Yardley made a huge contribution to American cryptology in establishing the first civilian agency, proving the concept of practical codebreaking and, above all, showing what good signals intelligence could achieve.

Yardley's turbulent life also exposed dilemmas that still run through the intelligence world, not least the ethics and legal basis of intruding on privacy, and the limits of what can or should be made public about intelligence work itself.

6

Tolkien Misses
the Cut

In the bowels of the Doughnut in Cheltenham are GCHQ's archives, including a treasure trove of intelligence reports stretching back over the decades. But there is one area that is protected by a large cage, access to which is even more restricted than to the main archives. It does not contain nuclear codes or the identity of JFK's assassin. Instead, it has the personnel records of living staff. There is of course a practical reason for this: if you are going to persuade recruits to be vetted and to sacrifice some of the freedoms of workers in the non-secret world, they have to be assured that you will protect their personal information indefinitely. But the cage is also symbolic. Despite all the brilliant technology, it is the people who are most precious to the organisation. One of the people whose file is not there is J. R. R. Tolkien. Understanding why he didn't make the cut tells us something about the creation of this strange enterprise.

At the start of the Government Code & Cypher School's existence, after the chaotic experience of Room 40, the kind of people that should be recruited became the preoccupation of Alastair Denniston, all the more so as he observed the world moving towards another global war. How staffing decisions were made illustrates the skill of predicting the needs of ill-defined tasks in an uncertain future. Building a new organisation out of a group of people, some eagerly taken on board, some who were simply available and others who turned up by luck, posed a unique challenge.

The formal beginning of GC&CS is generally taken as November 1919, when a cabinet committee decided to continue at least some of the work and methods that had been developed, despite the natural euphoria of peacetime. The official purpose of GC&CS was 'to advise as to the security of codes and ciphers used by all Government departments and to assist in their provision'. In short, it was to be an early version of a cyber security agency, protecting government information. Its parallel secret remit was 'to study the methods of cipher communications used by foreign powers' – in short, to break the diplomatic codes of other countries. This dual mission, defensive and offensive, runs through to modern GCHQ, although in the early years the defensive aspects were given less attention.

The new GC&CS, which combined Room 40 and MI1b, retained some thirty of their personnel, who effectively became the first professional codebreakers. Denniston was appointed as head, at the insistence of the Admiralty, but ownership of GC&CS transferred before long to the Foreign Office, a logical move since both the navy and the army had all but lost interest, as well as funding. Radio monitoring was reduced to a minimum but the volume of diplomatic cable traffic rose continuously, and a remarkably brief section of the 1920 Official Secrets Act gave the government complete power to acquire the communications of cables from all international cable companies.

Denniston moved his team at first into Watergate House off the Strand. This was suitably close to the Savoy Grill, a favourite of Admiral Hugh Sinclair, Hall's successor as director of naval intelligence and the new head of the Secret Intelligence Service (MI6), under whose oversight GC&CS operated, albeit as a distinct organisation (Sinclair later brought the Savoy Grill chef with him to Bletchley Park). The residents of Watergate House defied categorisation and included Leslie Lambert – better known as the society magician and short-story writer A. J. Alan – as the only radio expert, and Ernst Fetterlein, the chief crypta-nalyst of the last tsar, Nicholas II. Fetterlein had defected after the Russian Revolution and was quickly able to read the new revolutionary government's communications.

Throughout the 1920s, GC&CS read encrypted messages from almost every country it wished, including the newly improved American codes introduced in time for the Washington Naval Conference in 1922. More broadly, GC&CS was proving its worth identifying Russian subversive activity in Britain and Japanese strategic intentions. But it was not making much head-way with Germany.

By the end of the inter-war period Denniston was realistic about GC&CS's capabilities. With its very small staff it could not read and translate every telegram, and deciding which to focus on depended very much on its customers in the Foreign Office. But he was proud of the organisation's ability to break almost anything when necessary: 'We reached 1939 with a full knowledge of all of the methods evolved and with the ability to read all diplomatic communications of all powers' (with the notable exceptions of Russia and Germany).

Most importantly for the future of Bletchley and GCHQ, Denniston thought about recruitment as the international clouds started gathering again in the mid-1930s. He knew something of what was needed from the Room 40 experience and the successes of the 1920s, and he was also watching the technological developments in cryptography. In 1926 he sent his deputy, Edward Travis, to Berlin to buy a new machine in commercial use with the brand name 'Enigma'. It cost £28 and was arguably the best investment ever made by GCHQ (it is still on display in the museum at the Doughnut).

Acquiring the machine at that early stage did not lead to breaking it easily – that is a much longer story – but it demonstrated that encryption was becoming automated, mechanised and electronically driven. Codebreaking was moving out of the realm of linguistics, so the kind of people needed to tackle it would have to change. While there had indeed been some limited mathematical expertise in Room 40, a general prejudice remained in favour of the arts and classics over maths, but it seems that Denniston instinctively realised that the kind of puzzle-solvers he would need should include people with a more scientific approach to problems, to complement the linguists.

Recruiting from the hotbeds of Communism

Denniston's decision to approach the universities now seems an obvious one, but it was not so self-evident in the 1930s, with both Oxford – whose Union had passed the famous motion in 1933 that 'This House will under no circumstances fight for its

King and country' – and Cambridge being regarded as hotbeds of pacifism, Communism and appeasement. Denniston, however, discreetly used the network of trustworthy academic contacts from Room 40, including Frank Birch, who had left a Fellowship at King's College to pursue a career as an actor, including a much-lauded run as Widow Twankey in *Aladdin* at the Lyric, Hammersmith.

At a time when it was assumed that reliability could be determined primarily by the old school tie, communist sympathisers like the public school-educated Guy Burgess, Kim Philby and Donald Maclean had started working for Soviet intelligence, so Denniston's recruitment drive had to be based not only on the core skills he needed but also on the wisdom and experience of those doing the search. This was all the more important because the work was too secret to be described, and the imminence of war was not to be discussed. But, as he later said, 'there were men now in senior positions who had worked in our ranks during 1914–18. These men knew the type required.'

Recruiting people to jobs that cannot be described in an organisation that is secret remains a challenge for GCHQ and other agencies. One of those approached in the spring of 1937, E. R. P. Vincent, professor of Italian at Cambridge, later described the approach made to him by Frank Adcock, a Fellow of King's who had worked in Room 40 and went on to join Bletchley. After a long dinner, Adcock checked outside his door in a manner that looked like something from a spy novel and then told Vincent

that he was authorised to offer me a post in an organisation
working under the Foreign Office, but which was so secret
that he couldn't tell me anything about it. I thought that if
that was the case then he need not have been so cautious
about eavesdropping, but I didn't say so.

From the mid-1930s Denniston drew up lists of possible
recruits 'of the professor type', who were given brief introduc-
tory courses on cryptanalysis at the headquarters shared by
GC&CS and SIS near St James's Park tube station. These lists
still exist, with pencilled comments – possibly Denniston's – on
the candidates who attended the courses.

Intriguingly, one of those listed is J. R. R. Tolkien – described
as 'keen' and specialising in Scandinavian languages – but he
was not selected after his training course. Tolkien enthusiasts
who find it hard to accept that he was not wanted have suggested
that 'keen' is there to guide pronunciation of his name, though
comments on other candidates ('good') make this less convinc-
ing. It may simply be that he was not the type needed. There is
a certain irony here – Tolkien, along with other writers of heroic
fantasy fiction, is hugely popular in GCHQ, and entrance tests
in the distant past have included texts in the Elvish languages of
Middle Earth.

Those who were selected and ready to report for duty at the
outbreak of war were still predominantly modern linguists and
classicists, but two mathematicians were included: Gordon
Welchman, an established Cambridge maths don, and the somewhat younger Alan Turing, a brilliant Fellow of King's who had
just returned from a spell at Princeton. Even if Denniston had

done nothing else, simply recruiting these two was to change both the course of the war and, in Turing's case, the course of technology, computer science and much else.

Recruiting for jobs that had not yet been created to solve problems that had not been defined meant – in 21st-century terms – looking for 'aptitude'. Given the endless permutations involved, it was not surprising that proficiency at chess was seen as a possible indicator of suitability, along with a knack for completing the *Daily Telegraph* crossword. It is sometimes thought that Denniston's thinking about the types he needed was as vague as this. But from his private correspondence, including letters to civil service paymasters, it is clear that he understood the shift towards mechanisation in cryptography. Within a few years of buying the Enigma machine, he had already concluded that there were new areas where

> *a knowledge of modern languages appears to be of little account … the advent of mechanical means of cyphering has opened up a new phase of the work [of GC&CS] and as it develops it is taxing our resources in men capable of conducting investigations on machines to test the security alleged for them by their inventors.*

Recruitment at Bletchley also meant taking risks. Giving the responsibility for solving an enormous existential problem to a group of mostly young people – some of them with their undergraduate studies still not completed, others, in time, just school leavers – was a considerable gamble. Pointing them in the right direction, giving them what was needed and managing the anarchic interplay between each other – and between them and the Establishment – were to be Denniston's chief tasks. He did not always succeed, but his model of self-effacing leadership (he never describes his own achievements in Room 40 or elsewhere), and his focus on recruiting and enabling those who might not have been able to work in any normal bureaucracy, or work together at all, set a pattern for GCHQ that has paid dividends over the subsequent decades.

But for all this planning, there was also random selection, especially after the war began and Bletchley started to function – after the able-bodied were conscripted, Denniston had to make do with what was left. One of the many quotes attributed to Churchill about Bletchley staff has him saying to the head of MI6, 'I know I told you to leave no stone unturned to find the necessary staff, but I didn't mean you to take me so literally.' It seems highly unlikely that he actually said this, but it may well have been a self-deprecating joke by staff about their own strange stage, on which some remarkable characters played by chance, and sometimes quite by accident.

Practical cats and seaweed

Enter Geoffrey Tandy, a keen amateur dramatist, who earned his living as a marine biologist. Born in 1900, he came from a small Worcestershire village where his parents ran the pub. He progressed via Kidderminster Grammar School to studying forestry and moving on to marine biology in the 1920s. He eventually secured a steady, salaried job as assistant keeper of botany at the Natural History Museum in South Kensington, where his papers are still kept.

Although he enlisted for the last year of the First World War in the Royal Field Artillery, Tandy's service was limited, and his passion for sailing led him to the Royal Naval Reserve. In fact, his interests and enthusiasms outside work, from poetry and sailing to cats, were more important to him than marine biology and at least as formative. He and his first wife Doris became close friends of T. S. Eliot, and he worked as a broadcaster with Stephen Spender, W. H. Auden and Benjamin Britten among others. It was in Tandy's soft Worcestershire accent that Eliot's *Old Possum's Book of Practical Cats* was first read on the BBC Home Service in 1937, two years before its publication. His interest in cats and his large collection of feline photos may even have been part of the poet's inspiration.

He had been most engaged in the late 1920s and early 1930s, when he took part in expeditions to the Great Barrier Reef and the Dry Tortugas islands in the Gulf of Mexico, both trips combined his love of sailing with the study of coral reefs and algae. By the outbreak of war in 1939, Tandy had risen to become head curator of botany at the museum, but his heart was not in it and he was getting bored by the academic pursuit of natural history in London.

At around this time the War Office, under instruction from Alastair Denniston, had been searching for anyone who understood cryptograms and puzzles, while, as mentioned, Bletchley itself began quietly using the *Daily Telegraph* crossword as a way of finding problem-solvers on a larger scale. Tandy was not an obvious candidate. But a typographical error by the War

Office resulted in calls for experts on cryptogams, rather than cryptograms, to report urgently to Bletchley Park for unspecified duties. Cryptogams, of course, have nothing to do with puzzles or codes but are a class of seedless algae and ferns, best described to those of us who are not biologists as seaweed.

If nothing else, Tandy was something of an expert on seaweed and so it was that he took himself to Bletchley Park and swapped the relaxed world of dried plant specimens for blood-curdling briefings on the sensitivity of the code-breaking project underway there. He was told that if he broke the Official Secrets Act he would be shot. It says something about the secrecy of the place, and perhaps the etiquette of the times, that Tandy was too polite to ask why he had been summoned, and the staff at Bletchley never questioned why he had been sent. He spent the rest of the war there.

We know a little about what he actually did. He was head of Naval Section VI, whose job was not to break codes but to build a kind of library of 'equivalents' for the cryptographers. When an intercepted message contained a technical or engineering term that was not in any existing dictionary, NS VI would scour instruction manuals and official documents stolen or recovered from German, Italian or Japanese forces. It was vital but unglamorous work. We have an account by Carmen Blacker, who later became a distinguished Japanese scholar, relating the sheer tedium of cataloguing endless military terms that meant as little to her in English as they did in Japanese. But the importance for the military was clear, and timely use of technical translations like these was critical during the D-Day landings.

The story also goes that Tandy's particular background and expertise were not wasted. Despite his familiarity with broadcasting, we have no record of his taking part in the regular amateur dramatic performances at Bletchley, but his knowledge of marine biology was not irrelevant. At critical moments over the next three years, Enigma codebooks were recovered from German U-boats, sometimes at great personal cost. Waterlogged and permeated by salt, these documents needed salvaging, preserving and maintaining. Who better to consult than a recognised authority on saltwater algae?

Tandy himself comes across in the rare descriptions we have of him as a slightly autocratic and driven figure, determined never to let any member of staff leave his section, whatever the reason. We can only guess at the strain he may have been under, but his friend T. S. Eliot attributed his breakdown around 1950 to his experiences during the war. He worked for GCHQ for a while before retiring and resuming his theatrical interests, which his children have also pursued.

Geoffrey Tandy was neither a genius nor obviously suited to a job that had not been invented when he arrived at Bletchley Park and about which he could not possibly have known in advance. Whether he really came as a result of a typographical error, or whether this was a Bletchley Park intellectual joke told about him, is impossible to say. But his colleagues thought it a good story.

The central truth is that Tandy, along with so many at Bletchley, some of whom were carefully selected and others somewhat randomly drafted in by the three military services, on the face of it lacked the skills necessary to play even a small part in the greatest codebreaking, data-processing and machine-building experiment of modern times. Likewise, he was ill-experienced in leadership and never became a good manager. His greatest skill, however, was his understanding of how to catalogue and cross-reference information, drawn from all those years sifting through plant specimens in South Kensington. No modern human resource process would have picked out him or many of his contemporaries at Bletchley. And yet it worked.

Tandy would probably not have made it through recruitment into a modern high-performing company, yet he was a valuable part of the mix and constructed a particular service that was essential to the success of the whole effort. He sums up the amalgam of oddness, random skills and dedication that led to spectacular innovations and sustained success in the secret world of intelligence.

Faced with a challenge that even the greatest experts at Bletchley initially regarded as probably impossible, this ragbag of disparate talent somehow created something so extraordinary – the processing of data on a huge scale and eventually by a 'computer' – that we are still enjoying the benefits long after the threat of invasion has passed into history.

After the war and in recent years, the rationale behind selection to the secret world has not always been any more coherent. I once discussed the extraordinary crop of recruits at Bletchley with Roger Penrose, Nobel Prize winner and one of the great scientific polymaths of the last hundred years. He had been taught by, or worked with, most of the mathematicians involved, whose names will fit into place in the next few chapters. In later life Max Newman, the force behind Colossus, the first programmable electronic computer, constructed at Bletchley Park, became his stepfather. Through Newman he met many more of those who had moved on to GCHQ at Cheltenham after the war.

We cannot now interview these young men and women who made up the core of the codebreaking enterprise, and very few left any written reflections on their own aptitude as recruits. But Penrose gives some hints, and he himself exemplifies many of the qualities that these academics brought with them.

As we will see, the desire to solve puzzles and play difficult games runs through their work. Penrose as a young man was fascinated by visual patterns and puzzles. At school he constructed a dodecahedron and a tetrahedron out of Perspex with a hacksaw, and challenged other students to put them together. Later he worked on 'impossible shapes', notably the Penrose triangle, on which the artist M. C. Escher based some of his most famous lithographs of water flowing uphill and never-ending staircases. Penrose tiling – an infinite symmetrical pattern that never repeats itself – is commonly seen in modern design.

This fascination with puzzles and 'impossibility' runs through the stories of Bletchley Park and GCHQ. In Penrose's case it is inseparable from his remarkable insights and research, including his work with Stephen Hawking and his Nobel Prize-winning conclusions on how black holes are formed. For the codebreakers and their modern descendants, puzzling is a continuum that goes from the recreational to the deeply serious.

While mathematics is often seen as key to Bletchley and GCHQ's success, in practice familiarity with some of its foundational concepts was more important than any particular expertise. As one famous codebreaker put it, mathematicians 'tended to be good at this sort of work', but maths was not essential. At the heart of this was a fascination with probability. Penrose recalls his frustration as a child at being repeatedly beaten at the rock paper scissors game by his older brother Jonathan (later a chess grandmaster). Penrose went away and worked out a basic algorithm for the possibilities. This did not make him win all the time, but his new-found insights into probability and the psychology of his opponent stopped him continuously losing. Since then, advanced computer algorithms have been created to try to narrow down the probabilities involved in this ancient Chinese game.

Probability was at the heart of the challenge at Bletchley Park and remains so at GCHQ. Faced with the millions of possibilities churned out by an Enigma machine, the first task was to reduce these to a manageable volume. A whole range of techniques might be employed to do this, from capitalising on the mistakes of human operators to understanding the frequency of distribution of letters in any language. In particular, the theorem of the Rev. Thomas Bayes underpinned Alan Turing's approach. Bayes's insight was that a mathematical formula could calculate probability based on prior knowledge and developing information, meaning that the likelihood of something occurring could be refined and updated constantly. Turing's young assistant, Jack Good, of whom we will hear more, became an authority on Bayesian statistics after the war, a use of probability now much favoured by governments and companies as a guide to decision making in everything from public health policy to space travel.

Looking at the requirements for strategic thinking, practical employment of probability, and fascination with games and puzzles, it is not surprising that there was and is a close correlation between chess and codebreaking skills. But not all great codebreakers were particularly good at chess, nor were they all quick-thinking. Reflecting on Turing's reputation for being a slow and deliberative thinker, Roger Penrose commented that he himself had been very bad at mental arithmetic as a child. Having been demoted maths sets at school, he finally encountered a very enlightened teacher, who realised that Penrose would do extremely well if given a lot of time. By staying on in the classroom for another hour he could reach the 100 per cent that some of the quick thinkers would never reach, however much time they were given.

Many of those selected by Denniston would not have passed today's standard recruitment tests, which prioritise speed and practical focus. Much of this book explores these counter-intuitive choices and the extraordinary things they achieved together.

It is unclear why Tolkien did not make it into this group, nor why, a decade or so later, Penrose himself applied to GCHQ and suffered the same fate, something that still bemuses him. Reading his interviewer's notes upside down across the desk (a useful skill for spies and civil servants), he gleaned that a military vetting officer had vetoed his recruitment. Beyond a possible black mark from his headmaster, there was no obvious explanation. The truth is as likely to be the result of chance as much as anything more deliberate; the chaotic approach to recruitment unfortunately works both ways. Different choices might have deprived us of the *Lord of the Rings* trilogy, and Penrose also concedes that it would have been difficult to pursue his chosen studies had he been accepted in the secret world; 'Things,' he concluded, 'have turned out all right.'

7

Orchestrating
the Chaos

You may not be surprised to learn that in an organisation with a heavy presence of engineers and techies, discussions about how best to shape the internal structure of GCHQ often gravitated not to examples from the Harvard Business School, but from *Star Wars*, Hogwarts and Ernst Stavro Blofeld's SPECTRE. These instincts date back to the *Alice in Wonderland* world of Room 40 in the First World War and the extraordinary arrangement at Bletchley Park, which did away with traditional hierarchies. How this innovative start-up was moulded into the world's first big data-processing organisation has useful contemporary lessons.

The science fiction and fantasy examples teach some valuable lessons. Organisations that rely on top-down autocracy, such as the Galactic Empire in *Star Wars* and SPECTRE in Bond, display a poor track record in creativity and innovation. In the world of global baddies this is partly because personal terror is not a great long-term motivator; Darth Vader is a less than empathetic line manager and Blofeld seems to enjoy the gruesome death of his own staff just as much as the pursuit of his enemies. But notwithstanding their personal styles and limited emotional intelligence, the bigger problem is that their leadership and organisational structures dictate the kind of outdated solutions that such enterprises produce. SPECTRE, typically based in a hollowed-out volcano or on a remote island, is obsessed with vast undertakings that are immensely complex and never finished quite quickly enough. The same is true of the two 'Death Stars' in the *Star Wars* series. The megalomaniacs of the Empire do not seem to learn that big, centrally driven projects almost always have a single point of failure. Repeating the project is likely to bring the same results and, while you can sympathise with his motivation, Darth Vader's solution of physically strangling the project managers simply exacerbates the problem.

Hogwarts, on the other hand, looks a little more like the secret world of Bletchley and GCHQ. Dumbledore's powers are ill-defined and he presides over a good deal of chaos. He clearly has authority but it is largely exercised through wise influence rather than command and control. The various fiefdoms of the professors at Hogwarts are regularly in tension with each other and allowed to go their own creative, often erratic ways. If there is a grand plan from the centre, it is not always clear; the books chronicle endlessly creative attempts to see off an ever more obsessive Voldemort, who, to be fair, has a more subtle and asymmetric approach to conquering the world than the military models of Darth Vader or Blofeld. At Hogwarts, there is an overarching goal, but everything else is delegated.

At first glance, Bletchley Park may have looked chaotic. It certainly did to Joseph Eachus, who arrived in England in July 1942 as the first formal US navy liaison officer to this new part of the British secret world. He had been given minimal training for his mission ('They showed me where the passport office was') and on arrival asked the first head of GC&CS, Alastair Denniston, for an organisation chart to help him understand the place. 'I don't believe we have one,' replied Denniston. 'I didn't pursue this with him,' Eachus recalled, 'but I was never quite sure whether he meant we don't have a chart, or we don't have an organisation.'

Eachus was not a man fazed by chaos. At the outbreak of war he had just finished his PhD in physics and was teaching at Purdue University, Indiana, the primary educational power-house for engineering in the United States at the time. On responding to a government questionnaire and listing one of his hobbies as cryptanalysis, he was offered a correspondence course on the subject. He found himself commissioned as an officer in the navy's OP-20-G, the cryptanalysis unit, his military training consisting of a quick lesson in how to salute and not much else. According to an official record,

On his first day in Washington he walked into the office to find one man talking to himself, trying to learn Japanese; another was sending Morse code to himself and a third was on the phone, trying to get a second phone on his desk to ring by routing the call through Alaska.

Eachus immediately felt at home, and he spent the rest of his career in signals intelligence in the navy, the US National Security Agency and later in the computer industry.

The culture of Bletchley Park therefore came as no shock, and he soon discovered that access to US navy supplies of sugar and coffee, rationed and scarce in Britain, helped build his relationships. While working on the technical solutions to decrypting German naval Enigma he met Barbara Abernethy, a young linguist who was among the first to join at Bletchley and had moved from codebreaking to administration, as Denniston's personal assistant. They married after the war.

Their liaison was one of many, a reminder that the romantic view of Bletchley Park as a place run by a small group of chaotic eccentrics who cracked everything on their own is not entirely incorrect. There was at least plenty of romance.

But the interesting puzzle is this: how did an organisation variously described by contemporaries as 'rudderless', an 'asylum' and 'creative anarchy' become a vastly successful factory of intelligence? By D-Day, Bletchley Park employed some 10,000 staff (with arguably twice that number in associated organisations around the world) and was providing Allied commanders with the most comprehensive picture of what the enemy was doing and thinking that has ever been available to wartime leaders.

In the process, Bletchley's staff created machines that could process great volumes of data at high speed and ultimately a computer – Colossus – that was programmable. This extraordinary achievement, under incredible pressure and sub-optimal conditions, is a puzzle worth investigating. How was this chaos organised to produce something amazing? To begin to answer this, we need a brief digression into the birth of the organisation and what its production line – the 'intelligence cycle' – looked like.

Captain Ridley's shooting party

As we have seen, Bletchley did not spring up from nowhere. While Britain was chronically underprepared for war in 1939, Denniston was among those who had seen conflict as inevitable and done his best to prepare at least his contribution to the national effort, assembling a remarkable collection of individuals to work for him. For his part, Admiral Sir Hugh 'Quex' Sinclair, the head of MI6, had already bought a rather gloomy and architecturally underwhelming gothic mansion outside the Buckinghamshire village of Bletchley as the venue for wider intelligence efforts, including some MI6 work and GC&CS. Bletchley Park was selected not so much because it was on the train line between Oxford and Cambridge, or in easy reach of London, although both were true; its main advantage was good 'connectivity' in 20th-century terms, being well served by telephone cable lines. For a signals intelligence organisation this was and is the thing that really matters. When a site was sought for GCHQ after the war, Cheltenham was chosen because the US forces stationed there had developed excellent cable connections to London and beyond, which far outstripped accessibility by road and rail (and still do).

The small group of GC&CS staff who arrived to survey the site in 1938 were given the unlikely cover story of being 'Captain Ridley's shooting party', as Ridley was one of the administrators responsible for the move. By 1939 the old hands from Room 40 and GC&CS's inter-war years had been supplemented by the new university recruits who had been through a basic cryptanalysis training course, and that year saw the now famous 'Huts' rapidly thrown up to house the different sections that dealt with all the diverse problems. Denniston fought and won a succession of battles with his bosses to keep the organisation on one site, creating a unified campus model that continues to be a key feature of GCHQ's success and a challenge as it expands geographically.

Denniston and the early joiners faced two fundamental issues: the complexity and volume of data, which remain the core contemporary challenges both for GCHQ and for many businesses operating in the digital era. In short, the introduction of electro-mechanical encryption machines such as Enigma presented problems that many of those involved thought might never be solved. Despite some successes against the early version of Engima during the Spanish Civil War, the models in use at the start of the war by different branches of the German military had been improved in ways that vastly increased the complexity of the task.

Alongside this complicated intellectual challenge was the overwhelming problem of the scale, volume and organisation of data. From the huge quantity of radio transmissions across the Axis powers, stretching from the Atlantic to Japan, to the number of Enigma and other machines in use (some 35,000 by the end of the war), GC&CS was confronted by a tidal wave of data, much of which was unreadable. Even if a breakthrough could be made against a particular cipher on a particular day, how could this be scaled up? How could whatever was deciphered be delivered to the relevant parts of the Allied military in time for it to be useful?

Moreover, how could any of this be achieved while preserving absolute secrecy, because the knowledge that communications were being intercepted would certainly lead to the Axis powers improving their security? From the very first moment, in fact built into the job of codebreaking, was a sense of paranoia and precariousness; at any moment the whole edifice could collapse. Codebreakers are in a race not only to decrypt material today, but to make sure they stay ahead of the next improvement in the enemy's technology tomorrow. That in essence remains the contest in which GCHQ is engaged against modern adversaries, from geopolitical threats to terrorists and cybercriminals. But, as we shall see, this is not merely a description of the organisation's role – it is part of what drives it to perform at its best. Being at the cutting edge means being close to falling off the edge, and it provides an irresistible challenge.

In short, Bletchley was being asked to do something that might not only be technically impossible, but also to handle, process and distribute unprecedented volumes of data at speed, while at the same time concealing the fact that it was being done at all. Success for Bletchley would, at best, be a permanently fragile enterprise, always on the point of failing and 'going dark'.

Faced with a task of such epic proportions, on which so much was depending in the struggle for national survival, and without a clear definition of the problem set, let alone the exact skills required to meet it, Denniston brought together old thinking and new and disparate disciplines. Alongside the mathematicians and cryptanalysts, the military and civil service administrators and support staff, there were information-processing experts from the British Tabulating Machine Company and engineers from the General Post Office. In fact, there was an increasingly bewildering array of skills and experience – including staff from department stores and manufacturing – collected across a growing temporary estate around the original mansion, each with its own culture and way of operating.

To exacerbate Denniston's problems, some of the staff were his, but most were military (to be more accurate, largely conscripted civilians rather than career military personnel), and lines of accountability were complex and often blurred. Individuals were recruited and paid in very different ways, and few people at Bletchley had only one boss.

The great military historian Ronald Lewin once described Bletchley as a 'honeycomb' organisation, but that implies a uniformity across its separate cells. In practice Bletchley was an eclectic arrangement of units, large and small, some run on the novel lines of Dilly Knox's all-female team, others operating as factories. It was the mathematician Gordon Welchman who saw the need for a production line – in fact a workflow process – that would bring all these disparate enterprises together into a single coherent effort. Even before Enigma was being broken, he had worked out how to do this.

The orchestration of these many and varied parts into a coherent whole, with multiple lines of control and accountability, and different cultures but a single mission, was the template for modern-day GCHQ, and it offers some wisdom for other organisations. Although dysfunctional at times – perhaps all the time – and alongside the brilliance and intellectual camaraderie came plenty of sheer drudgery and boredom, Bletchley was ultimately successful.

It is worth examining what enabled the 'organisation' of Bletchley Park to function, and to do so it is important to keep in mind what the intelligence factory actually did. There were a number of distinct stages in the production line.

First, messages were intercepted at Y stations within Britain and in locations around the world. As the war progressed, new stations were opened in many new places, including mobile units moving with the army. Their positions – whether in Australia, Egypt, India, Malta or Singapore (until it fell) – depended largely on physics: where could radio transmissions be best heard?

Most were abandoned after the war but some, like Brora on the east coast of Sutherland in northern Scotland, remained active through the Cold War, monitoring Soviet naval traffic until 1986. Staff remembered the beautiful scenery and the warmth of the local community, but Brora was at times a bleak posting. As at any Y station, the work was often unrelenting and monotonous, with interceptors noting down Morse code from crackling transmissions that came and went. Accuracy in recording streams of largely unintelligible data was critically important; getting Morse wrong made the task of Welchman's traffic analysts who were trying to decrypt the message even more difficult, if not impossible.

These stations might be owned and operated by one of the three branches of the armed forces, or by an array of other organisations, from the Metropolitan Police to the General Post Office. Staffed predominantly by young women, once the messages had been intercepted, they had to transmit them to Bletchley, either by motorbike or by teleprinter link.

The material then entered a second stage at Bletchley itself, being registered and allocated to the relevant section – German naval messages going to Hut 8, army and Luftwaffe messages to Hut 6, and so on. Alongside decryption there was a wealth of detail to collect on the traffic itself, all of it meticulously recorded on index cards for cross-referencing.

Successful decryption led to an intelligible German text, which then – in the penultimate stage – had to be translated. This was not straightforward, and particularly in the early days linguists who were used to reading Goethe and Heine struggled to make sense of German military, technical and scientific language, some of which would have been unintelligible to them even in English. As time went on, Bletchley accumulated a vast store of this detail, much of it in Geoffrey Tandy's section, which acted as an explanatory encyclopaedia for analysts to consult.

Finally, another group 'sanitised' the material, removing anything that made it obvious that it came from a direct, intercepted radio message. It had to look as though the intelligence might have come from human spies within the German war machine, run by a fictional master-spy called 'Boniface'.

Once the material had been analysed and its importance judged, it had to be disseminated to the War Office, the Admiralty and the Air Ministry, as appropriate; in addition, Bletchley's Special Liaison Units, set up by MI6, were established with key commanders around the world, allowing quick and easy transmission on the spot by people who understood the importance of the material and the absolute necessity of protecting it.

Although there were particular procedures for very urgent messages, all of this needed to be done at speed if the material was to be of use.

This 'intelligence cycle' illustrates some of the particular tensions at Bletchley Park. At the start of the war, the army, navy and air force generally assumed that the job of Bletchley was simply to decrypt and translate; everything else should be done by the service professionals. They also each regarded their own service as the priority, especially when their personnel were being used. With notable exceptions, there was a general scepticism towards and distrust of Bletchley's group of civilians and amateurs.

At the heart of these turf wars – unremarkable to anyone who has worked in any large organisation – was a disagreement about the fundamental role of Bletchley Park. What was its task? Was it just 'translation'? What the GC&CS leadership had illustrated was that to be effective, and to keep everything secret, Bletchley Park had to be an integrated intelligence operation. 'Signals intelligence' required those involved to understand the technology behind the signals, how they were created, transmitted and intercepted, as well as the nature of the people – their adversaries – operating the network. Tackling Enigma and other mechanised encryption required not just cryptanalysts, but data processors and machine engineers, administrators and subject-matter experts, and it required these disciplines to mesh together in new ways.

This was an internal battle that Bletchley Park ultimately won, but it was a hard grind. A major reorganisation in 1942 settled the arguments, and also led to the effective removal of

Alastair Denniston to London. He was replaced by Edward Travis, a more effective organiser and a forceful leader, although without Denniston's collegial approach. Denniston's health was poor and three years of war had taken their toll. He was not a man who enjoyed administration but he had applied himself to it; his private correspondence shows that he had done so with much greater insight into how people and organisations work than those around him assumed.

The culture that Denniston encouraged and Travis preserved was highly distinctive. To get the best out of the young recruits who formed the core of the codebreaking huts, military hierarchy would largely need to be abandoned. This was not simply to give them a warm and familiar university feeling, but because a meritocratic approach was the best way to deal with the complexity of the codebreaking process. In the research sections it would also create an atmosphere in which innovation and progressive ideas could flourish.

When Bill Bundy arrived as the senior US liaison officer in Hut 6, he and his group were welcomed and briefed by Gordon Welchman. According to Bundy, it was agreed that, irrespective of nationality, each 'was to be judged on his merits, regardless of rank, no matter what this did to the structure'. In practice this was already happening: 'a major working for a civilian alongside a sergeant, whoever was the smarter of the two'.

In this part of the Bletchley culture, there were few formal commands: 'orders were nearly always given in the form of requests and accompanied by explanations'. In Welchman's Hut 6, perhaps the pinnacle of the Bletchley Park way of working, the internal history records that:

the guiding principle all along was not to lay things down from on high, but to bring everybody into consultation, to get general agreement and to make everybody feel participants and not cogs in an unintelligible machine.

After the war, Welchman reflected on how GCHQ should be structured. In a secret note he made suggestions that at the time looked revolutionary, and even now run against the grain of rigid civil service and corporate rules. The dominant theme was that 'people with particularly valuable skills should be able to reach the top levels both of salary and prestige without having to perform tasks for which they might not be particularly well qualified'. He suggested that it should be possible for a brilliant individual 'to be drawing a higher salary than the Director without having to undertake any administrative duties whatsoever'. Against a civil service and corporate prejudice that pegs importance, seniority and pay to the number of staff a person manages, and very often their time served, he argued that 'promotion and prestige in an organisation such as GCHQ ... should depend more on an individual's value to the organisation than on the number of his subordinates'. Modern GCHQ, constantly torn between the secret and civil service worlds, has made steps in this direction, valuing technical talent in financial and status terms.

In Hut 6, the mathematician Hugh Alexander's fellow international chess player and lifelong friend Stuart Milner-Barry had been a reluctant stockbroker at the outbreak of war. Chess was his passion – both men, and Harry Golombek, another chess master, had been in Argentina at the Chess Olympiad when war was declared. Milner-Barry became an enthusiastic manager, and he describes a culture of delegation, where competence was assumed and reporting to superiors minimal, that would be out of the question to most modern organisations. Hut 6 was:

a very loose and informal organisation with only an indispensable minimum of formal routine meetings. I left heads of department with a very free hand ... they did the same with their head of subsections, and heads of subsections with their head of shift. Each department would have its own meetings ... whenever it felt like it. When major changes threatened, we would all get together at all levels, but avoiding as far as possible the monster general gatherings at which it is almost impossible to get anything done.

Milner-Barry became a good leader, developing what he described as the 'aimless wandering' approach to management, where he would 'go round and talk, or perhaps encourage other people to do the talking and listen ... one can learn more in this way than any number of formal interviews'.

The very small administrative team around the head of GC&CS did not attempt to impose central structures or procedures on these units. Each area was allowed to govern itself on the basis of what worked. Crises – and there were a few – were dealt with, sometimes rather slowly, but there was little central micromanagement. Issues of prioritisation or personality disputes were generally sorted out locally.

The author of the secret official history of Hut 3, one of a number of histories produced immediately after the war, described the staff's 'exceptional freedom ... those who did their work well were left, within the inevitable limits, to do it their own way'. This was 'the exact reverse of the Hitler principle of the greatest possible meddling with the greatest possible number'.

The key to the high levels of performance was a personal sense of responsibility:

If mistakes were made, as of course they were, by ignorance or negligence, the remedy was found not nearly so much in reprimands, or witch-hunts for the delinquent, as in the mortification decent persons feel at having let things down.

Failure in any intelligence organisation, whether a message was not decrypted in time at Bletchley or a terrorist attack plan is missed at a modern agency, leads to an acute sense of frustration and disappointment. It may be that nothing more could have been done, or that something could, but either way it is the opposite of what the agency is trying to achieve and is experienced as a defeat by all those involved.

In Max Newman's section, which created the computer known as Colossus, Peter Hilton recalled, '[Newman] realised that he could get the best out of us by trusting to our own good intentions and our strong motivation and he made the things always as informal as possible.'

But this pre-Silicon Valley culture was not uniformly imposed either centrally or through peer pressure. Mr Freeborn of the British Tabulating Machine Company, for example, always wore a business suit and was addressed as 'Mr Freeborn'. He had a reputation for being in total command of the detail of his section and was fearsome to those unprepared; Welchman later recalled that the wiser academics learnt how to consult him for his expertise rather than try to dictate requirements.

'Something to do with biscuits'

Very occasionally, when disputes became so acrimonious that they threatened the functioning or security of the whole, as happened at one point in Hut 3 during 1941, personnel were removed. An internal memo to Denniston describes one naval officer as 'interfering, intriguing, creating and magnifying difficulties and misunderstandings, causing friction, undermining confidence and, incidentally, making proper liaison impossible'. Empire building and rigid allegiance to service hierarchies ran counter to everything Bletchley was trying to do. In the end, and after too much delay, this long-running dispute was resolved by the introduction of a remarkable Macclesfield businessman as head of Hut 3.

Eric Jones was neither a cryptanalyst nor a mathematician and spoke no German, leaving school at fourteen and becoming an executive in the textile industry. Peter Calvocoressi, who worked with him, perhaps reflected a certain snobbery in Bletchley's culture when recalling that 'he was believed to have had something to do with biscuits' in the Midlands. And even when Jones became director of GCHQ in the 1950s, holding the helm through the Suez Crisis, he was sometimes referred to as the 'Manchester businessman'. As with a number of other post-war GCHQ leaders he had not been to university – unthinkable for a civil service grandee in Whitehall or for the heads of MI5 or MI6 – but his external experience was exactly what was needed. Travis noted that 'his even temper and calmness never fail to exert a deflationary influence upon a highly strung team'. He exercised a 'benign despotism' and, while respecting his subordinates, he was not intimidated by their intellectual skills.

There was method to Jones's leadership beyond his personal qualities. Having witnessed the problem of inter-service rivalry, he took active steps to educate Hut 3 in its single mission, which meant compromising on the 'compartmentalisation' necessary for security. He organised lectures in quieter periods so that the various sections could understand what others did; he mirrored the practice of Hugh Alexander in Hut 8, who boosted morale 'by giving everyone the fullest and most current information about the results obtained from the work'. His blend of motivation and firm control worked: perhaps more than any other individual, he was responsible for turning Hut 3 into the efficient, fast-paced factory of intelligence that was so critical to supporting the D-Day landings.

Importantly, different parts of the organisation were very different in feel and leadership. In the Typex room, where Wrens (WRNS, the Women's Royal Naval Service) processed up to fifteen million letter groupings each month, there needed to be a more industrial approach, with Bletchley Park's internal management papers recording that:

It became necessary to … institute factory methods. This was done chiefly by keeping records of output per watch, per machine and per girl. This showed up weaknesses, peak hours etc., and enabled the manager to adjust numbers and skills per watch.

This was not an organisation without metrics, where they mattered.

There was very considerable drudgery, low morale and sickness in some of these areas. Incessant noise, poor ventilation and the sheer monotony of the work took their toll. It is a reminder, for all the romantic and high-octane stories of codebreaking, that much of the intelligence work and data mining was, and still is, painstaking and dull. What united those sections that operated with freedom and those arranged in factory style was the sheer hard work over long hours.

At its best, Bletchley Park's culture enabled a flow of information within the Huts. This was critical to success, and enabled senior people to resolve the inevitable problems of prioritisation, such as the use of scarce 'bombe' machines or procurement of new machinery. It helped that many of them were friends and most lacked egos, starting with Denniston, a striking contrast to some of the inter-service and government machinations. Denniston regarded it as part of his job to manage Whitehall and absorb the flapping of others in government, allowing his people to get on with the job – a task familiar to his successors as heads of GC&CS and GCHQ. In fact, Denniston was an unlikely template for modern leadership, someone who could tolerate difference and preside over tribes and individuals in creative tension with each other.

The organisational achievement and scaling-up of Bletchley Park can best be seen by the changes in its provision of food. The formal seated banqueting arrangements provided by Admiral Sinclair and his Savoy Grill chef at the tatty mansion in 1939 had by 1944 been replaced by a cafeteria supplying 32,000 meals a week, covering twenty-four-hour shifts, seven days a week. Reflecting wartime rationing, staff verdicts on the food inevitably deteriorated from 'wonderful' in the early days to official complaints that the cafeteria served 'mince, which is chopped gristle and slime, and fishcakes, which are made of mashed bones and breadcrumbs'. The exception was Christmas parties, when the chef's surprising access to the black market seemed to help. In fact, Bletchley knew how to party on a shoe-string, creating bunting and Christmas decorations from sliced-up code books (a fragment of which we presented to Queen Elizabeth II when she opened the new National Cyber Security Centre).

For all the administrative headaches and bureaucratic frustrations resulting from expansion – including reassuringly modern notes from Denniston pleading with staff to bring their cups back – Bletchley maintained the kernel of its mission and its workforce throughout the war and onwards to GCHQ. Similar frustrations, including missing crockery and unacceptable noise levels, remain, but the much more fragile ecosystem of creativity, innovation and high performance is what really matters.

The bizarre organisational structure of Bletchley Park was summed up in a note by Nigel de Grey in 1943. A veteran of Room 40 (and the 'Dormouse' in the Lewis Carroll parody), de

Grey was a key figure at Bletchley and left a lengthy account of its administration full of minute detail, along with sometimes acerbic comments. He observed and noted everything, from the widespread mental health problems among staff to his endless fights with Whitehall. He describes a start-up chaos that not only worked, but excelled.

I suppose that if you were to put forward a scheme of organization for any service which laid down as its basis that it would take a lot of men and women from civil life and dress some of them in one kind of clothes and some of them in another, and told all those dressed in black that they came under one set of rules and all those dressed in white under another and so on, and then told them that they had a double allegiance, firstly to the ruler of their black or white or motley party and secondly to another man who would partly rule over all of them, but only partly, any ordinary tribunal would order you to take a rest cure in an asylum. But suppose that the tribunal were somehow foolish enough to adopt your idea and in order that you might begin your work said 'We will now lend you some tools – they may not be quite what you want but you must make do with them, and tell us when they get blunt and we'll see if we can sharpen them for you', some higher power would presumably lock up the tribunal as a public menace – or, if it were in Russia or Germany, shoot them out of hand. Yet that is in fact the precise organization of Bletchley Park. Now it happens that Bletchley Park has been successful – so successful that it has supplied information on every conceivable subject from the movement of a single mine sweeper to the strategy of a campaign and a Christian name of a wireless operator to the introduction of a secret weapon.

8

Triangular Pegs in Triangular Holes

If you had to write an HR manual for the early years of GCHQ it would be titled somewhere between 'Making the best of what you've ended up with' and 'Getting difficult and brilliant people to move in the same direction'. Faced with the selection of staff, random and planned, the challenge for the leadership was – and still is – to allow them to flourish. Three snapshots illustrate how the first and second heads of GC&CS managed this. They concern Dilly Knox (he of the warm baths), Gordon Welchman and Max Newman. The results of managing these people successfully were not trivial: the contribution of intelligence to the military victory of D-Day, the foundational approach to data and knowledge management, and the creation of the world's first digital computer. In the process came a redefinition of what codebreaking really means.

An all-female team

In August 1939 Dilly Knox installed himself in one of the cottages beside the mansion at Bletchley Park. At that early stage, Alan Turing was in the loft, where food would be sent up to him in a basket by rope pulley to enable him to work undisturbed. Nominally the chief cryptanalyst, given his Room 40 experience and his knowledge of Enigma, Knox was unsuited to stepping back and directing operations, large or small, and he had never run a group.

In those initial days his primary focus, alongside Turing and Peter Twinn, the first mathematician to join GC&CS (and, after the war, an authority on the longhorn beetle), was Enigma. In the period 1939 to 1940 this included a critical dialogue with French intelligence cryptanalysts and, most importantly, with mathematicians from Polish intelligence. At a key meeting of experts from the three nations in the Pyry Forest south of Warsaw, where the Polish General Staff's Cipher Bureau was located, Alastair Denniston and Knox heard the extent to which the Poles were ahead; they had already successfully decrypted a significant number of German messages. Much of the progress had been reversed by German improvements to Enigma but, crucially, the mathematicians had developed an electro-mechanical machine to help them, which they called the *bomba kryptologiczna*.

After the outbreak of war, Turing travelled to Paris to meet the exiled Polish mathematician Marian Rejewski and his colleagues to take forward discussions of the machine. This meeting in turn led Turing to develop the British bombe, which was constructed and installed at Bletchley by the time the Luftwaffe's campaign against British cities began. The bombe did not decode anything, but it crunched through the possible machine settings of Enigma until it found a match, at a speed no human could manage.

This was the most important foreign partnership in GC&CS's early history, although soon to be dwarfed by the 'special relationship' with the United States. Denniston had appreciated the technical help that allies could bring by sharing progress and ideas, an approach he later pioneered in the key alliance with American cryptologists. He realised very early that signals intelligence was a team activity in which the involvement of other countries – with their access to communications in different locations and their particular technical specialisations – was essential. That continues to be true – 'sigint' is by far the most collaborative area of international intelligence, including its role in cyber security; the technical and mathematical problems are such that a scientific style of sharing information is essential, and the scale of the work is too great for any single country, even a superpower.

Knox's contribution to this international phase of the project was not easy. Never a diplomat and incapable of masking what he thought, he was already a sick man, having been diagnosed with lymphatic cancer. He had been particularly bad-tempered at the three-nation meeting, and Denniston had to take the unusual step of writing to his French counterpart Gustave Bertrand to apologise for him:

He is a man of exceptional intelligence, but he does not know the word cooperation. You must surely have noticed that off duty he is a pleasant chap loved by all. But in the office his behaviour is different. He wants to do everything himself. He does not know how to explain anything. He can't stand it when someone knows more than him. Unfortunately I cannot do without him, he knows more about the machine than anyone in the country.

Denniston's solution was gradually to allow Turing and others to focus on large-scale mechanical solutions while letting Knox get on with what he did best, researching his own subset of difficult problems. This gentle reorganisation – and the proposed role for Knox – led to one of Knox's many resignation threats. As always, Denniston replied calmly, pointing out:

If you design a super Rolls-Royce, that is no reason why you should drive the thing yourself up to the house, especially if you are not a very good driver ... Do you want to be the inventor or the car driver? You are Knox, a scholar with a European reputation ... The exploitation of your results can be left to others so long as there are new fields for you to explore.

In among the justified flattery, there may have been a subtle dig at Knox's terrible reputation for driving. One of his unfortunate passengers recalled that 'he had a remarkable theory that the best way to avoid accidents was to take every crossroad at maximum speed'.

The 'new fields' for Knox to explore turned out to be an inspired move. For this he required new staff, and he initially recruited an all-female unit, including Mavis Lever and Margaret Rock. While this group was patronisingly referred to as 'Dilly's girls', Knox himself was wholly focused on qualifications and had learnt that the 'mathematical input to cryptography' was crucial, emphasising to recruiters that 'he did not want any debutantes whose daddies had got them into Bletchley through knowing someone in the Foreign Office'.

Both came from London universities, Rock a mathematician at Bedford College and Lever a modern linguist at University College. They and their colleagues were fiercely loyal to Knox, and he grew into an inspirational and generous leader of his group, constantly giving credit to his team for codebreaking success, and ensuring that they were recognised and thanked by the leadership at Bletchley.

Two of those reasons for gratitude to this remarkable unit of young women stand out. In the very early morning of 29 March 1941 the director of naval intelligence, Admiral John Godfrey, rang through to Bletchley: 'Tell Dilly we have had a great victory in the Mediterranean and it is entirely due to him and his girls.' The Battle of Cape Matapan, off the south-west coast of the Peloponnese, which had taken place over the preceding forty-eight hours, was a much-needed tonic for the British public at a low point in the course of the war, and Churchill used it to great effect in bolstering morale.

The destruction of an entire cruiser division of the Italian navy constituted its most significant defeat of the war, undermining its ability to fight at night and to engage in any further large-scale confrontation with British and Australian ships of the Mediterranean Fleet under Admiral Sir Andrew Cunningham.

The sequence of events preceding the battle encapsulated the challenges of the Bletchley intelligence cycle, illustrating some of the problems that still persist for modern agencies trying to pre-empt an attack. A decoded message from Luftwaffe units deployed with Rommel's Afrika Korps indicated that German aircraft were being moved to Sicily to support an unspecified naval action. Cunningham, based at Alexandria, needed precise details in order to cover the large distances in the Mediterranean and intercept the Italian fleet. Knox focused his team on this task and on 25 March they broke a signal that read 'Today is X minus 3'.

For Mavis Lever and her colleagues, an anxious wait followed for more detailed intercepted radio transmissions brought from the Y stations by motorbike despatch riders. The key intercepts were not direct instructions to the Italian fleet, but were repeats or copies of those messages sent to a commander in Rhodes.

At this point, the secrecy of Bletchley's capability became the key issue: how to allow Cunningham to use the information without betraying its origin. Dissemination of Ultra material (the codeword used to cover this most sensitive source) was arranged through a series of liaison officers around the world, whose job it was to get close to key commanders and ensure they had the most sensitive intelligence, while protecting the way it was described and used so as not to reveal where it came from. This is a system that still operates, and it has been used in Afghanistan and any conflict where GCHQ supports the military.

In the case of Matapan, the cover story was to be that a reconnaissance plane had spotted the Italian fleet by chance. Having seen the intelligence, Cunningham told his own staff about the supposed reconnaissance flight and judged that the Italian intention was to attack Allied troop convoys on their way from Alexandria to Greece. Having ordered his fleet to prepare discreetly to sail after dark, he ostentatiously went ashore for a game of golf, making sure that he was seen by the Japanese consul while talking loudly about a dinner party later that night. He then returned quietly to HMS *Warspite* and sailed off to engage the Italians.

The reconnaissance flight cover story was briefed to the press after the battle. The usual recriminations followed, with the Italian and German militaries blaming each other, neither imagining that Enigma was breakable. Such is the power of conspiracy theories that stories were being published as late as the 1980s attributing the events to an alluring spy called Cynthia, who had seduced the Italian naval attaché in Washington and persuaded him to hand over the Italian naval cipher.

The true version was celebrated in a long poem in Victorian style written by Knox, listing the ten women in his team who had been responsible. Significantly, it included those who provided administrative support (wisely placed there by Denniston to keep Knox functioning) and the security-cleared tea lady, making no distinction between the importance of their respective contributions.

There was no conflict between the deeply serious work in hand and the flippant approach to codebreaking as a 'game'. For Knox, whose section became known as ISK ('Illicit Services

Knox'), games and puzzles did not constitute a separate field, reflecting his approach to making sense of the world. Sometimes the strange thinking patterns that went on in his head popped out into the external world – one morning at Bletchley he greeted Mavis Lever and Margaret Rock with huge excitement at the door to the cottage: 'If two cows are crossing the road, there must be a point where there is only one, and that is what we must find.' Among his favourite recruitment questions was 'Which way do the hands of a clock go round?' Anyone who answered 'Clockwise' would be told, 'Not if you are inside the clock.' Modern tech company recruiters did not invent these odd tests of aptitude, but the difference with Knox was that he was personally fascinated by why the answer should be so.

It was up to the women in his unit to work out what he meant as he threw out ideas that he frequently could not articulate in immediately comprehensible terms. Mavis Lever was clear about the role of Knox's team:

Bright ideas sparked off like Catherine Wheels and Margaret Rock might be trying to pin down the latest one while Mavis Lever was still struggling with yesterday's, which might be a winner and might have been forgotten in the excitement of today's.

In the case of the cows crossing the road, Knox was referring to his progress in breaking the Enigma used by the Abwehr, German military intelligence, the second great advance made by his section. Despite being a flawed, corrupt and disjointed organisation, the Abwehr had global reach and could shed light on German perceptions of almost any part of the world. From intercepting this traffic and, more importantly, analysing German police messages, Bletchley could see into operations in the territories conquered in eastern Europe and began to see references to the systematic persecution of Jews.

But most urgently, Bletchley could see what the Abwehr thought of the material coming from its network of agents in Britain. Since this network was almost completely fake and under the control of MI5 – the famous 'double cross' operation well chronicled by Ben Macintyre – Bletchley could monitor whether the misinformation was being believed. This assumed great strategic significance as D-Day approached and MI5 mounted the most impressive deception operation in military history. 'Garbo', the agent whom the Abwehr regarded as their crown jewel, along with his fictional network around the country, was critical to impressing upon the German High Command the belief that the invasion would be in the Pas-de-Calais instead of Normandy. The intricate and elaborate 'intelligence picture' conveyed to the Abwehr was enthusiastically received, and it was largely thanks to Knox that Bletchley knew that this deception had been read and believed by Hitler himself.

As deteriorating health forced Knox to hand over control to his protégé Peter Twinn, ISK grew to a hundred staff, with a more formal system in place (including two bank managers as registrars) while preserving the core spirit of research. By the end of the war some 140,000 Abwehr messages had been decoded by this group.

A 'normal' organisation would have found it difficult to deal with Knox's temperament, his difficulties with other people, and a less patient and thoughtful leader than Denniston would certainly have despaired of him. More importantly, a coherent, streamlined modern organisation would have seen his inability to adapt to mechanisation and the mathematical approach to cryptanalysis as a fatal flaw, running against the tide of progress. His poor administrative ability and lack of interest in the management of people and processes would have sidelined him, or worse, in any contemporary structure.

But significantly, the mathematicians never doubted Knox's ability and value. The great strength of Bletchley's organised chaos was that it allowed Knox, and others in a range of specialist and idiosyncratic areas, to function in their own space and their own way, and to excel. Denniston's personal triumph was not merely to placate a difficult man when he threatened to resign, as Knox regularly did, but to push back when necessary in order to manage the tension with other sections. Had Knox been completely isolated it would have negated the impact of his work, so the challenge was to link his team to the rest of the organisation in a way that worked for both sides. That explains in part the bizarre wiring diagram of Bletchley that Joseph Eachus asked for on his arrival but never received.

The success of this approach was in the ultimate output from Knox. Without his knowledge of Enigma before the war, progress would have been slower in tackling the new variations of the machine introduced from 1939. Without his team's breaking of Italian naval Enigma, there would have been no Matapan and the Italian navy would not have been effectively eliminated from the course of the war. But above all, without his work on the Abwehr, Britain would not have had the insight into German intelligence that was critical to the success of the double cross operation and the subsequent invasion of Normandy. Knox died of cancer before witnessing the achievements of D-Day, in which his odd section had played such a key role.

Data about data about data

One of the many staff who had misinterpreted Knox's silence for disapproval when he first arrived at Bletchley was Gordon Welchman. A teacher of algebraic geometry, he knew nothing about codes:

> *I was absolutely green, and I simply tried to learn what I needed to know as quickly as possible. I was interested in finding out what the problem was, not in how it had arisen.*

Welchman went on to make two huge contributions to the effort against Enigma. First, and perhaps more than anyone else, he was responsible for the scaling-up of Bletchley's capacity to deal with the ever-increasing volume of data. He realised that it needed to become a factory production line of information, and saw earlier than most the practical benefits that would flow from mechanising the process and introducing an industrial-scale system of knowledge management and intelligence production.

Second, he discovered that the content of a message was not necessarily the most important thing, and that a whole world of information was available simply by looking at the outside of messages and the way they travelled. So-called 'traffic analysis' became a key foundational concept for GCHQ, especially through the Cold War when Soviet codes were often impenetrable. In the cyber world this data about data is known as 'metadata' – the information that directs and identifies emails, for example, as they are delivered around the world. It is a sophisticated digital version of the outside of an envelope – the content is secondary. Doing clever things on a vast scale with this data about data was an extraordinary breakthrough and arguably one of the most precious secrets of Bletchley Park – something that needed to be more closely guarded over the next forty or more years than many of the details of breaking Enigma itself.

Welchman's contributions came about almost by accident. As he later recalled, he was relegated to an empty building with one other colleague and a constantly growing pile of unreadable Engima messages, and asked by Knox to look at 'call-signs and discriminants'. At first he assumed this was a way of side-lining him.

In order for any radio message to be efficiently transmitted it needed some unencrypted information as a preamble. The sender had to communicate a range of details, including their own call-sign and the destination, the time, the number of letters in the message – and a three-letter 'discriminant'. This indicated which key on the Enigma machine the receiver should use, because different parts of the German military and state apparatus used different keys, even if they were using the same machine. In addition, the British intercepting station added some further data of their own on time of interception, radio frequency used and other relevant information.

Welchman and his colleague sat down to the unglamorous task of looking at these technical preambles to messages and followed an approach that he had used before with problems. He would start writing and see what emerged: 'We started making lists of this and that and gradually we came across things that seemed of interest.'

Welchman's view of codebreaking suddenly changed. He had previously believed 'the common view that cryptanalysis was a matter of dealing with individual messages, of solving intricate puzzles and of working in a secluded backroom, with little contact with the outside world'. Thanks to Knox's instruction to work in just such a secluded room, he now discovered quite the opposite. The data about the messages he was looking at, although he could not read them, revealed an entire communications system, technical and human, providing a glimpse into the lives of his adversaries and exposing a world of structures in which people had to organise their communications. There was a formality, including military politeness, about the messages, enabling Welchman to begin to imagine and then reconstruct these networks.

Welchman spent his early months getting to know the Y station radio interceptors, who were a great source of experience: simply by listening to radio transmissions all day and night they began to recognise voices and patterns, and even particular styles of Morse code keying. As he expanded his team, Welchman and his assistants Patricia Newman and Peggy Taylor began the painstaking work of creating lists and diagrams, using coloured pencils to represent the different discriminants for the various arms of the German military.

This traffic analysis, as Welchman came to describe it, had been used in the First World War in a limited way, notably to spot large-scale German naval activity. But Welchman saw the value of doing it on an industrial scale. What began as a task focused on the problem of breaking Enigma itself soon developed into a methodology that conferred two great benefits. Just as in simpler times, when the address and origin of a telegram guided the censor in deciding what was worth reading, so the data around the Enigma messages helped prioritise and select what was worth decrypting. It therefore assisted in addressing the problem of volume, since even at the height of its codebreaking powers, Bletchley could never hope to decrypt more than a fraction of what was intercepted around the world.

Second, traffic analysis could reveal incredibly valuable information without anyone ever decrypting the content of messages. Each of the many networks was given a colour as a key name. One of the colours of the discriminants – Brown – was the least exciting-sounding network, signifying the Luftwaffe research unit. But this was a regiment that trialled new radio technology and, because it was not high-status, practised some very careless security, including easily guessable codewords and the encryption of much of its traffic with systems well below Enigma's sophistication.

Bletchley collected, catalogued and organised vast quantities of seemingly trivial and ephemeral information on the personalities, call signs and geographical locations of the Brown radio network. This information was card-indexed so that it could be cross-referenced and searched. Despite the wartime paper shortages and the challenges of logistics, the Park was consuming two million Hollerith punch cards per week – six tonnes of paper. The indexing system was dull to operate but as crucial to code-breaking as anything going on across the site, constituting the collective memory of the Bletchley Park brain.

The indexes of Bletchley Park are a dry subject but worthy of study by anyone interested in knowledge management. Thousands of intercepts flowed into the Huts in no particular order and very often appeared to contain nothing of interest. But an unidentified name of a lowly German officer, stored away and cross-referenced, might prove critical when it cropped up again in a different context.

Peter Calvocoressi, who became head of the air section in Hut 3, described the Air Index as the 'central repository of what Ultra knew about the Luftwaffe. Its importance cannot be exaggerated.' The index was devised by Reggie Cullingham, who had worked at Kelly's Directory, a kind of Victorian 'Yellow Pages' (for readers under fifty, a large thin-paper version of a website such as Yelp). Cullingham was a fine example of one of the many talented individuals at the Park who possessed a unique skill:

From first to last the Air Index was designed and nurtured by a strange genius, a man of everyday appearance and attainments who had, and he knew he had, a particular gift for that very thing: indexing. He was a triangular peg in a triangular hole.

The many indexes at Bletchley were so precious to the whole enterprise that a copy of the 'Central Index', into which all the different indexes were collated, was kept in the Bodleian Library in Oxford, in case the original was ever damaged or destroyed in a bombing raid.

This was an early indication of what could be achieved through the bulk mining of data, without any access to the content of the message. At its heart was a painstaking understanding of the enemy as a network of human beings trying to operate a communications system; it was as much about people as machines.

An internal report from 1945, released in 2018, stated that 'all the information on the performance of [the] V1 Flying Bombs [...] before the operational launchings was derived from low-grade cipher and TA [traffic analysis]' by the Brown network. Given the danger V1s posed to London, this was quite a tribute.

In the decades that followed, traffic analysis played a key role in GCHQ's success in the Cold War, and during the modern era analysis of bulk metadata has become a critical component of counter-terrorism and cyber security.

Welchman was not a cryptanalyst but he understood the principles behind the machines Turing and the engineers at Bletchley were inventing. His personal improvement to Turing's bombe design was dramatic, reducing the time taken to find the Enigma settings by many hours. His lasting achievement, however, was to imagine the bigger picture of producing intelligence at scale, perhaps precisely because he was not buried in specific problems. The insight of how much traffic analysis, in bulk, could reveal about the people behind the communications, was an amazing breakthrough in intelligence work.

Tea parties: the birth of digital programmable computing

The final stage in Bletchley Park's evolution is best represented by Max Newman, Alan Turing's mentor at Cambridge and life-long friend. Newman was a relatively elderly forty-five-year-old and an established Fellow of the Royal Society when he arrived in 1942. After trying a number of areas for research he persuaded Edward Travis, who had by now taken over from Denniston, to allow him to attempt something truly novel.

By 1942, Y stations had been intercepting traffic for two years that was neither in Morse nor generated by Enigma. This network of communication began to grow steadily across German-occupied Europe, and with success against Enigma becoming more reliable, attention turned to cracking 'Fish', as the network had been nicknamed, and particularly the traffic on the German armed forces network known, along with the machine that encrypted it, as 'Tunny', another name for tuna (inventing groups of names at Bletchley followed a process that will be familiar to viewers of *Only Connect*).

The Tunny transmissions were encrypted by Lorenz machines attached to teleprinters, a method that was much more conven-ient for the user than Enigma, as essentially you typed in a lengthy message and it was typed out at the other end. The process used with Enigma machines of encrypting each letter and copying down the results was done away with, along with much – but not all – of the potential for human error. The level of encryption too was significantly higher than that in use with Enigma; critically, however, no one had actually seen a Lorenz machine and it was not until 1945 that one was captured. It was clear, nonetheless, that Lorenz was being used for some of the most important communications of the Nazi regime, including personal messages and orders from Hitler.

John Tiltman, the veteran of MI1b, had made some progress in decoding this traffic, and a huge breakthrough came from a very young mathematician called Bill Tutte. Both were helped by a basic mistake made by a Lorenz operator who re-transmitted the same long message with minor differences, giving the codebreakers something to go on. Tutte effectively reverse-engineered the Lorenz machine using pure mathematical theory.

Max Newman's insight was that the manual methods developed by Tutte to break the Tunny traffic could be mechanised, building on the concept that had proved successful with the bombes. Newman wanted to create a complex machine that would run two paper tapes simultaneously to test the Lorenz wheel settings. One tape would be a Tunny message, the other would be suggested wheel settings for the Lorenz. By running them at 2,000 letters per second, the machine could effectively match the right settings to the text.

The General Post Office engineering department constructed a machine that was nicknamed 'Heath Robinson' – a nod to its extraordinary look and noise, and perhaps the fact that, while effective, it regularly broke down or the tapes snapped, or both. A particularly brilliant General Post Office engineer, Tommy

Flowers, who had also worked with Turing on the bombes, helped build these new machines, although he was convinced that their mechanical workings could be improved by becoming fully electronic.

Flowers proposed using thermionic valves, essentially to automate electronically a process that would have been impossible to run manually. He was used to working with these valves because, as he later wrote, 'the need for superfast processing had arisen in the telephone industry in the years immediately before the war'. He went away and developed his own design, estimating it would take a year to develop. On 1 March 1943 Newman asked for money both for more 'Robinsons' and to help with Flowers' new project. It was typical of Newman, and an important lesson for GCHQ's approach to innovation, that both were allowed to go ahead, even though there was pressure on Newman to back only one project.

Since the full story became public, it has been generally agreed that Colossus I and II – huge machines the size of a room – were the world's first programmable computers, with Bletchley Park therefore standing at the very start of computer science. The vision that Turing had set out in his famous paper of 1936 – which earned him the reputation as the father of computing – was becoming a reality.

Whereas previous machines had been built for specific tasks, Colossus could be asked to do different things. While it did not have a stored memory, it pointed to the possibility of a universal machine that could change itself, ultimately learning to tackle different tasks. Something we now take for granted in a basic computer was becoming possible for the first time. After the war both Newman and Turing would pursue this concept, although without being able to speak publicly of Colossus. These machines were broken up, except for two that were taken to GCHQ to use against Soviet traffic.

A declassified 1945 report, written by those who worked on Colossus, captures some of the excitement generated by the world's first computer:

> *It is regretted that it is not possible to give an adequate idea of the fascination of a Colossus at work; its sheer bulk and apparent complexity; the fantastic speed of thin paper tape round the glittering pulleys; the childish pleasure ... the wizardry of purely mechanical decoding letter by letter (one novice thought she was being hoaxed); the uncanny action of the typewriter in printing the correct scores without and beyond human aid ... periods of eager expectation culminating in the sudden appearance of the longed-for score; and the strange rhythms characterizing every type of run.*

One of the authors was the mathematician Peter Hilton, who described the conditions that Newman created for his section: 'He realised he could get the best out of us by trusting to our own good intentions and our strong motivation and he made the thing always as informal as possible.'

Shaun Wylie, later GCHQ's chief mathematician, described Newman's efforts to make sure that innovation was not stifled in a shift system where people might not meet. Ideas could be left on blackboards or written in a book for discussion at regular tea parties: 'Anyone could call one, ideas were bandied about, you came if you could and brought your own tea. Tea parties didn't decide things but they led to action on all fronts.'

*   *   *

These three snapshots give a glimpse of the extraordinary achievements of Bletchley, illustrating some of the ways in which innovation and the leadership of staff were managed in particular areas. But they do not really come close to capturing the breadth of the organisation or indeed answering the question of whether it was organised at all. But it is fitting to start with remarkable people first and structure second. That has been the fundamental approach in GCHQ's secret world. Whether in Room 40, at Bletchley, through the Cold War and on to the internet and cyber age, innovation and creativity did not come from a pre-planned, top-down written strategy.

9

John Lewis and the 'Velvet-arsed Bastards'

One of the pleasant hazards of being director of GCHQ accompanies a visit to any supermarket in the Cheltenham area. Given the size of the workforce and the size of the town, the chances of running into a present or retired colleague are very high. One particular summer evening I had called into Waitrose on the way back from the Doughnut and noticed one of our mathematicians, who appeared to be counting the cans of soup on display. We chatted as if this were an entirely normal thing to do (in fact it turned out that he was preparing a challenge for a weekend treasure hunt), but he then talked about the chief cryptanalyst who had recruited him many years ago and his pre-war employment at the John Lewis Partnership. I subsequently discovered that the complex science of how food is displayed on the shelves of Waitrose and other supermarkets began with another GCHQ and Bletchley employee, and that a former John Lewis partner was at the heart of the most sensitive and longest-running counter-intelligence operation of the Cold War, which exposed the treachery of numerous British and American spies and only came to an end in the 1980s.

As I considered the connections between John Lewis, Bletchley Park and GCHQ, I began to see that this was about far more than the stories of a few exceptional individuals. John Lewis influenced the management approach that was a key part of some of the great technological advances at Bletchley and later GCHQ, and also contributed a significant number of women partners as data handlers. In return, GCHQ staff contributed to the research and retail systems that gave John Lewis and Waitrose a leading edge. Both possessed a cultural outlook that aspired to be meritocratic, non-hierarchical, innovative and progressive. In short, they shared common values to a surprising degree. Understanding how this happened tells us something about the successes and challenges of both enterprises, whether in the secret world or on the high street.

From social justice to chess

John Spedan Lewis's story has been told often and well. In 1909, at the age of twenty-four, a riding accident confined him to bed for two years, during which he reflected on his life and wealth. His realisation that his father, the founder of the family retail business, his brother and he himself each earned more annually than all their employees put together drove him to investigate profit-sharing and mutual business models. Out of this eventually came the John Lewis Partnership, co-owned by all the 'partners' or employees. Despite his father's reluctance, Spedan pursued this vision, not simply out of an attachment to social justice but for what he regarded as sound business sense. He also saw the mutual model as politically essential, the only progressive alternative to the rising tide of popular communism.

Spedan's view of leadership matched this mutual approach. He saw the senior leaders of the company as akin to Gladstone's cabinet, in which ministers had a high degree of autonomy and could express differences candidly, but would ultimately take collective responsibility. Spedan's was the decisive 'casting vote', yet he tried to avoid dictating policy.

He did, however, pursue a number of personal passions. One was the belief that hiring brilliant people without a fixed idea of exactly what they would do was a sound strategy for a dynamic and innovative business, an approach that was also tried at Bletchley; another was the conviction that research into emerging technologies would put the company ahead. These came together in his third passion: chess. And it was chess that introduced him to one of the great figures of GC&CS, GCHQ and 20th-century intelligence: Conel Hugh O'Donel Alexander, known as Hugh Alexander.

Alexander was born in Cork in 1909 and to the end of his life regarded himself as Irish, despite moving to the UK in his teens after his father's death. This identity as an outsider perhaps helps to explain his fierce independence and healthy scepticism towards authority. A brilliant mathematician, his devotion to chess prevented him from making the grade as an academic, but G. H. Hardy, one of the most famous mathematical celebrities of the last century, recognised his talent as a rare 'creative' mathematician. Instead, after graduating from Cambridge, Alexander became a schoolteacher at Winchester for six years, which allowed him time to develop his career in chess, and he eventually became an international master in 1950. This was a lifelong hobby – he published many books on the game and in his later years at GCHQ was also the chess columnist for *The Sunday Times*.

Spedan Lewis's amateur interest in chess brought him into contact with Alexander in the mid-1930s, beginning a friendship and a long correspondence that lasted until Lewis's death thirty years later. By 1938 Lewis had persuaded Alexander to join his retail business as head of research. He also had plans for a chess centre on top of the company's Oxford Street store, hiring the women's world champion Vera Menchik to run it, and he hoped to build a stable of great players.

The corporate 'research' job offered to Alexander came with a salary that was double his teacher's pay, and he constantly felt overpaid, regularly displaying what would now be called 'impostor syndrome'. Spedan addressed his friend's qualms of conscience, arguing that it was precisely because Alexander was not from the business world that he was valuable. He talked of his friend's 'clearness of mind, moral courage and sense of values', and 'a natural ability that is almost of the order of genius'. In the brief period before the outbreak of war, he deployed Alexander in a range of roles, including head of personnel, no small job in a business with 10,000 staff.

From this often uncomfortable experience, Alexander seems to have learnt three key approaches to getting the best out of people that were to be central to his work at Bletchley Park and later as head of cryptanalysis at GCHQ during the Cold War.

The first was a willingness to debate and discuss ideas widely. New and challenging ways of thinking were welcomed, as a glance at the John Lewis staff gazette shows. This was a cornucopia of internal news, external hobbies and hobby horses – political debate was not outlawed and radical proposals were welcomed, irrespective of hierarchy. At its best, the Bletchley of Hugh Alexander and Alan Turing, his first boss, reflected this approach. Even in the modern GCHQ, a lively internal social media site was similarly eclectic and sometimes impassioned, much to the nervousness of some managers. Topical and ethical concerns were not buried away but addressed and discussed at 'town halls', in the tradition of Bletchley meetings.

Explanation and shared enterprise, even in a secret environment, were key to Alexander's style, as he explained when reflecting on his experience of partnership at John Lewis:

> *The sort of things that make for efficiency in an operational cryptographic section are very much the same as in a business organisation. The idea is to get business efficiency whilst preserving the atmosphere of a cooperative voluntarily undertaken for its own sake.*

Second, and more importantly, alongside a general openness to discussion and challenge was a clear commitment to meritocracy. Spedan believed in this, even if in practice the principle was sometimes imperfectly applied. At Bletchley, where Alexander had been summoned in early 1940, it became central to the early codebreaking work, a remarkable achievement in itself in a military site during wartime.

Hut 6 was run by the mathematician Gordon Welchman, whom Spedan Lewis recruited as the new head of research at the end of the war on Alexander's recommendation. A now declassified official history, written at the end of the war, summarised what the senior US liaison officer Bill Bundy, who worked in Hut 6, described as the most complete meritocracy he had ever seen (he knew something about large organisations, having later served as a senior foreign policy official for presidents Kennedy and Johnson):

Over five hundred and fifty individuals of widely differing ages, gifts, characters, men and women, Service and civilian, British and American, yet formed with all their variety one welded whole ... and with less time wasted on intrigue than one could easily have thought possible.

To facilitate this, in many sections at Bletchley there was an informality in dress bordering on total scruffiness ('Who are these velvet-arsed bastards?' asked a visiting military chief as he surveyed the civilians in corduroy trousers). Long-haired twenty-year-old students might find themselves briefing very grand senior visitors, and even more remarkably for 1940s Britain, most staff addressed each other by their first name unless asked not to. These traditions have been enthusiastically upheld at GCHQ, with a dress code best described to me as 'unsmart casual', although with plenty of interesting and colourful exceptions. Long before the tech sector made it fashionable, GCHQ understood the relationship between creativity, culture and informality, and it is still a strength that senior visitors to Cheltenham from Whitehall may find themselves briefed by someone who looks like a teenager.

Finally, underpinning Alexander's discursive and meritocratic approach was an independence of thought and a willingness to challenge authority when necessary. It was he and Spedan Lewis who argued, against the consensus of senior managers, that conscientious objectors called up to non-military service should receive the same subsidy from the John Lewis Partnership as those fighting. Both were aware of committed Quakers in the company who were involved in exemplary non-military service. For Spedan it was about upholding 'the level of decency that is one of the chief glories of our own country', one that set it apart from what he called the 'blind ruthlessness' of Nazism.

It was Alexander, along with Welchman, Turing and Milner-Barry, who signed the famous personal letter to Churchill in October 1941, going over the heads of their bosses to plead for an end to delays in staffing. Alexander was unfailingly discreet about his secret work in his communication with Spedan, but in a letter from his digs at the Shoulder of Mutton Inn, Old Bletchley, he told his former employer of his deep frustration that the organisation was being run by a 'spineless dud' who would not 'kick up hell' to galvanise Whitehall.

In the same letter, Alexander asked Spedan a favour. With conscription about to be extended to single women aged between twenty and twenty-five, Alexander wanted to scoop up as many female John Lewis partners as possible to work on codebreaking. He knew their skills and also their familiarity with punched-card information-processing devices similar to the Hollerith machines in use at Bletchley. He was also acquainted with the commercial 'Colourdex' talent-management display system that the partnership had adopted along with the Admiralty and others, detailing the skills and career progression of each staff member. Within a few weeks he had successfully visited the head of personnel and was interviewing possible candidates.

And it was likewise Alexander, while a senior figure at GCHQ, who testified as a character witness for Alan Turing at his trial in 1952. Unfazed by standing up for a homosexual man in 1950s Britain, he described Turing as a 'national asset' and expressed his admiration for the moral courage he displayed in his handling of the trial.

After the war, Alexander agonised about returning to John Lewis but concluded that his talent lay in codebreaking, despite Spedan's best efforts to tempt him back. In the chaos of post-war Bletchley and its transition to Cheltenham as GCHQ, Alexander played a key role in establishing and running H Division, tasked with breaking Soviet codes, although most of his work remains highly classified.

He was also pivotal to the Venona project, one of the most far-reaching counter-intelligence operations in history. At the height of Bletchley Park's operations, their American counterpart based at Arlington Hall, a former girls' school in Virginia that served as the headquarters of US signals intelligence, had been amassing encrypted communications from Stalin's intelligence agencies, in the knowledge that the Soviet Union would not remain an ally for long. Between 1943 and 1980 American codebreakers at the National Security Agency, in close partnership with their British counterparts at GCHQ, began painstakingly to decrypt thousands of messages. Alexander was at the heart of this, overseeing the GCHQ effort and keeping up with the volume of material more junior colleagues were processing. It was from this most secret source – the details of which were allegedly withheld even from President Truman – that some of highest-profile double agents were unmasked. Kim Philby and the other members of the Cambridge Five; Klaus Fuchs and the spies in Oppenheimer's team at Los Alamos; and the Rosenbergs and other networks of Soviet spies in America were gradually identified and passed to MI5 and the FBI. Venona illustrated the brilliance of the codebreakers, the patience and triangulation of information required, and the dangers of secret material: not everyone mentioned in the messages was a Soviet agent, and treading a fine line between McCarthyite paranoia and complacency was a continuing challenge for analysts throughout the Cold War.

A new way of selling

Back at the John Lewis Partnership, Spedan was trying to rebuild his business, literally in the case of the bombed Oxford Street flagship store. Many partners had been killed during the conflict, including Vera Menchik, who died with her sister when their south London home was hit in an air raid. Spedan invested heavily in automation and found Gordon Welchman a very capable substitute for Alexander, and even better qualified to look at mechanisation. In the years before he left to start a new life in the US as a specialist in computers, Welchman seems to have prevented the partnership from wasting a great deal of money on computing projects.

But there was one final gift from the secret world to John Lewis, in the shape of a young and slightly bumptious student called Patrick Mahon. A weary John Lewis HR manager records an interview for the key position of buyer, to which this young man 'came up the stairs and into the room at a run', spoke 'almost without drawing breath' for ninety minutes and ran down again. By 1949 Mahon had been in the partnership for some years and had already achieved what his interviewer conceded were 'spectacular' sales figures.

Mahon had studied German and French before his degree was interrupted by the war. He arrived at Bletchley in 1941, at the age of twenty, and four years later became head of Hut 8, a successor to Turing and Alexander. The latter recommended him to Spedan, admitting that he had initially found Mahon hard to like, but that he had matured over the years. Mahon brought to the buying process at John Lewis the phenomenal work rate and organisational skills with which Alexander credited him. The intricacy of ordering, arranging supply chains and monitoring inventory to create one great retail symphony played to his passion for imposing structure on complexity. He became the key buying director for the partnership, and his radical approach – 'We need to introduce a "no sacred cows" initiative. [We] cannot surrender to entrenched conservatism' – reveals something of the Bletchley and GCHQ culture that shaped him, as did his call to colleagues to be 'receptive to new ideas and intolerant of people who just want to glide along to retirement in their well-established grooves'.

Mahon was briefly acting head of Waitrose and tipped as a future CEO of the John Lewis Partnership, before his early death in 1972. But his lasting contribution was to the science of supermarket sales. It appeared that Waitrose had very little idea of how to run a self-service supermarket and was still thinking along the lines of the then-current model of grocery shopping; this no longer involved personal service and simply arranged goods in some sort of logical order for customers to select. Mahon set about working out how to maximise sales by looking at likely footfall and traffic flows through the store. What seems obvious now – that customers need to be drawn through the shop and exposed to the widest range of goods – was novel at the time. Mahon understood that 'path planning' could greatly increase sales, and that shelf and display organisation made a measurable difference to the volume of business. If customers were directed towards high-value goods, this could yield a huge increase in revenue. By analysing customers' buying habits and the relative value of goods, Mahon transformed the John Lewis approach to self-service selling, a project that he regarded as the most daunting and exhausting of his life, including his years at Bletchley.

So the displays that my mathematician colleague and I were admiring on that summer evening in the Cheltenham branch of Waitrose were a strange and surprising legacy of the intelligence 'traffic analysis' that had made such a difference at Bletchley Park and throughout the Cold War. Central to this analytical approach was the culture of openness and meritocracy to which both organisations aspired. It is not insignificant that GCHQ was the only major intelligence agency, perhaps in any country, to be unionised, much to the irritation of Margaret Thatcher. Culturally, GCHQ was different from the hierarchical world of

Whitehall and closer to the model of mutuals or partnerships. This was the result not of an ideological or political agenda, but a conviction, born of Bletchley, about what actually works in the secret world of technology and innovation.

10

A Female Organisation

Prior to an age in which every organisation is striving to achieve gender equality, Bletchley Park represented a remarkable experiment. For reasons that were nobody's conscious choice, in one key respect this nascent GCHQ rapidly became chronically un-diverse. There were very few men, Bletchley being essentially a female organisation, where women made up 76 per cent of the workforce. And a chance encounter with a veteran one Saturday morning taught me that this was not just a statistic. It had a lot to do with what the place felt like – its culture – how the intelligence machine functioned and, ultimately, why it succeeded.

Women dominated every activity in every area of Bletchley except two: the guard force controlling access to the perimeter, and senior management. Since Bletchley was a remarkably un-hierarchical organisation, its management did not need to be large. Even in the smaller areas of technical codebreaking work, given the poor access young women had to mathematical education at university level in the 1930s, they were proportionately well represented.

Understanding something about what these women did, and why their achievement went uncelebrated and was arguably squandered after the war, tells us about pervasive attitudes to science, technology, engineering and mathematics, as well as wider social trends. It shows how, to a limited extent, Bletchley – and later GCHQ – was able to go against prevailing currents and in secret become a beacon of progress and the repository of at least part of the pioneering advances in computer science. But it is also the story of a failure both to recognise achievement and capitalise on technological prowess, with British primacy quickly being ceded to the emerging American computing industry.

Joan Clarke in her own right

Some years ago I was giving a talk at a girls' secondary school in Gloucestershire, a small part of a wider programme to try to increase the number of female engineering recruits at GCHQ.

As usual in these talks, I started with Bletchley Park, partly because *The Imitation Game*, the biopic of Alan Turing starring Benedict Cumberbatch and Keira Knightley, had been released the previous year. That seemed to provide an obvious entry point for this age group, and, for all its faults, the film had at least made an attempt to show a woman – in this case Joan Clarke – doing something vaguely significant at Bletchley.

Thanks to the film, Clarke has become well known as the close friend and fiancée, for a few months at least in 1941, of Alan Turing. She is rarely assessed in her own right – not least because her career after Bletchley was so highly classified – but Clarke's life was by any standards quietly exceptional. The daughter of a clergyman, she progressed from a London high school to Cambridge, where she achieved a double first in mathematics, although only in theory – women were not actually admitted as full members of the university and given degrees until after the war. One of her academic supervisors, Gordon Welchman, who was among the founding four mathematicians of Bletchley, recruited Clarke early in 1940.

Welchman told Clarke that the work she was going to do at Bletchley did not really require maths, 'but mathematicians tended to be good at it'. Clarke joined Hut 8, where Turing and others were responsible for tackling German naval Enigma ciphers. Breaking these was critical to the Battle of the Atlantic and the protection of Allied convoys from the US.

There is no doubting the importance of Clarke's contribution to Hut 8's work in keeping supply lines to the UK open. Hugh Alexander, who took over leadership of the section from Turing, regarded her as 'one of the best in Hut 8'. The titanic struggle for survival – and then dominance – in the Battle of the Atlantic has often been recorded. The facts are stark: by June 1941, German U-boats, operating in so-called 'wolf packs', were sinking 282,000 tonnes of Allied shipping per month. From July this fell to less than half that figure, and by November to 62,000 tonnes. The German navy's response was to introduce new codebooks and an enhanced version of Enigma. Hut 8 referred to this new cipher as 'Shark', and it was nine months before they could break it, during which period Allied losses mounted again.

The breakthrough came from some new analysis of weather reports and other transmissions but, in a reminder of the extended team exercise that defines codebreaking, this could not have happened without the capture of codebooks from a sinking U-boat off the coast of Egypt in October 1942. Before *U-559* sank, several men from the destroyer HMS *Petard* clambered on board to retrieve codebooks from the commanding officer's cabin and to try to remove cipher machines. Two of them, First Lieutenant Anthony Fasson and Able Seaman Colin Grazier drowned in the process, but not before handing the crucial Enigma material to a sixteen-year-old canteen assistant called Tommy Brown, acts of bravery that helped to change the course of the war in the Atlantic.

Joan Clarke was at the heart of Hut 8 throughout these dramas and, as one of a small rota of codebreakers operating a twenty-four-hour shift system, played a key role in breaking 'Shark'. With never more than 150 staff, Hut 8, later aided by the use of bombe machines belonging to the US Navy – an early example of the transatlantic intelligence partnership – broke over a million intercepted German naval messages over the course of the war.

After VE Day, Joan went to work at GCHQ until she married in 1952 and moved to Scotland. She came back to Cheltenham a decade later and picked up again with her fellow cryptanalysts until her official retirement in 1977 at the age of sixty, although she continued to do some secret work until the early 1980s. Her career during the Cold War remains classified but her role in the Falklands War has recently been revealed. Some forty years after she had broken German submarine messages, and long after her retirement, Clarke worked on and broke ciphers found on board the Argentine submarine *Sante Fe*, after it had been attacked in South Georgia in April 1982.

Once I had finished describing Joan Clarke's career at my talk in the Gloucestershire secondary school and noted that we had at long last named a room in the Doughnut after her, I spent a few minutes outlining the greatest technological breakthrough at Bletchley – the creation of Colossus, the world's first program-mable digital computer – which arguably led to the extraordinary computing power in the phones that my audience was politely pretending not to consult.

After the talk was over I could see two older women hanging back. As the sixth-form girls dispersed, the younger of the two came forward and told me that Joan Clarke was her aunt, while the elder of the two ladies said that she had worked on Colossus at the end of the war and knew Tommy Flowers, the General Post Office engineer who designed and built the computer. Before they melted away again into the school melee, we talked about life at Bletchley Park. My veteran friend began to describe what it felt like to work there, and I started to understand for the first time what it meant to say that Bletchley Park was predominantly female.

My assumptions about the role of women at Bletchley had been shaped by the popular histories that inevitably focused on the 'eccentric' men who were portrayed as driving the whole project. Women were seen either as socialites and debutantes doing secretarial work, or conscripted women in the navy (WRNS) or air force (WAAF) doing the drudgery of large-scale 'clerical' work. The truth is more complex and harder to assess, precisely because there are fewer records of what women actually did, despite the excellent work of a number of historians and the Bletchley Park Trust. Worse, the diaries and records we do have, both for women and men, tend to focus on social life outside work hours, for the obvious reason that staff were not allowed to write about secret intelligence. This in turn fed the idea that women were somehow lightweight debutantes and not an important part of the core team.

In fact, women were heavily involved in deeply technical work and what would now be seen as engineering, data processing and information management. With the exception of the significant number of women who moved on to GCHQ after the war, many of these skills were squandered in the post-war years.

A predominantly female workforce was also employed in the intelligence chain that fed Bletchley, starting with the Y service intercept stations. Without the collection of radio traffic at these sites around the world, and from the many other organisations intercepting enemy transmissions, there would have been nothing for the codebreakers to work on.

At the start of the war there was a scramble to get the right interception in the right places, and efforts could be somewhat basic. Daphne Humphrys recalled joining the Wrens, and after a rudimentary interview had established that she had a little German, which she did, and regularly tackled *The Times* crossword, which she didn't, she was posted to Kent. As Britain prepared to resist what looked like an inevitable cross-Channel invasion in 1940, Daphne and two colleagues were told to report to a deserted cottage on the cliffs to the east of Dover, where they would use a van equipped with VHF reception sets to monitor transmissions from German vessels at night.

Instructed to dress in civilian clothes, with 'the unbelievable cover … that we were factory girls on holiday and no one was to come within two miles of us', they spent long nights endlessly sweeping frequency bands looking for signals from German ships or U-boats. In the morning they would write up a report and walk two miles to the nearest post office to send it by registered post to the Admiralty. Since their van with its distinctive antennae could easily be seen through binoculars from the French coast, they were literally on the communications frontline and exposed to attack.

The network of Y stations, staffed almost exclusively by women, established itself as the fundamental building block of Bletchley Park's success. There was a vast amount of Morse code material to decrypt, but the Y stations also produced other kinds of intercepts of great importance – direction-finding signals, navigational beams, voice and teleprinter transmissions, and 'noise'. As Welchman had discovered on his early visit to an interception unit, the women listening to intercepted radio signals understood more about their adversaries than anyone else. As one young mathematician recalled, 'Some operators became so skilful that they could recognise their opposite numbers by their style of Morse transmission, just as one may recognise ... the touch of a well-known pianist.'

Accuracy and speed were extremely important, as mistakes could set back or even completely thwart the process of decryption. Most importantly, Bletchley and the Y stations needed to work seamlessly, prioritising the right traffic from an ocean of possibilities. And this could be traumatic work; Aileen Clayton records in her memoirs a night when she heard a German pilot, whose voice she knew very well, screaming for his mother as his aircraft fell towards the earth, causing Clayton to be physically

sick. Listening and watching in secret is still emotionally difficult; some of today's analysts – women and men – might
occasionally have to witness contemporary atrocities occurring
in the online world.

Back at Bletchley Park, it was certainly true that Denniston
regularly had requests to find work for the daughters of the
wealthy, and he sometimes obliged the parents of 'debs'. But
there was a rising group of female mathematicians and linguists,
reflecting the greater access of middle-class girls to education.
Sometimes their attainments were achieved despite their schooling; Ann Mitchell, who died in 2020, had studied higher maths
at school against the wishes of her headmistress, who felt it was
'not a ladylike subject'. She was one of only five women admitted to read mathematics at Oxford in 1940 and felt that she
lagged behind the men, 'who had been taught it properly at
school'. She worked in Hut 6 and her first job was 'to make
menus – a sort of program – on the machines. I don't suppose
we had ever heard the word "computer".'

Mitchell went on to become head of shift, a particularly critical
job that started after midnight when German codes were changed
and the daily race to break them commenced. She also worked
on 'duds' – messages that came out of the process as nonsense,
often because the German operator had got the machine settings
wrong or simply copied down the message incorrectly. There
were thousands of these errors, and solving them required a
mixture of maths and intelligent guesswork about what mistakes
a careless or tired operator might have made.

By contrast, others had been encouraged by more progressive
schools, chosen by their parents because of the opportunities

they offered for women. The Girls' Public Day School Trust (now The Girls' Day School Trust), a large network of schools founded in 1872 by the sisters Maria Grey and Emily Shirreff to provide an academic education for girls of all classes comparable to that provided for their brothers, produced a number of well-known suffragists. Girls sent to Trust schools included a number of codebreakers, among them Clara Spurling in the First World War and Margaret Rock in Dilly Knox's team. Rock is a prime example of how perception of female education had changed. We have moving letters from her father, a naval officer killed in the First World War, exhorting her to work hard at school and be academically ambitious. She was, and Knox described her as among the top four of the Enigma team and 'quite as useful as some of the "Professors"'.

As well as Joan Clarke, there were female cryptanalysts, probably more than have been recognised to date. Christopher Morris, a mathematician working on naval Enigma, recorded that some of the best cryptanalysts in the naval section were women. Given that only 16 per cent of mathematics undergraduates at Cambridge in the five years before the war were female, and add to that all the other obstacles to the recruitment of women, including much lower pay, the numbers were surprisingly high. Cryptanalysts constituted, however, a very small part of Bletchley, albeit a critical one, and most of the female and male graduates were employed on more general analytical or machine work.

Early computer programmers

As the task of processing huge volumes of data at Bletchley grew, more women were drafted in. The Newmanry – Max Newman's machine codebreaking operation – was the heart of the new technology, from the 'Heath Robinson' machines to Colossus I and II. What started as a section with one cryptanalyst, two engineers and sixteen Wrens, grew by April 1945 to twenty-six cryptanalysts, twenty-eight engineers, and two hundred and seventy-three Wrens. The secret report on Bletchley written at the end of the war described the training involved, and noted that some of the Wrens showed an advanced ability in cryptography and engineering, despite not being mathematicians. They were arguably blazing a trail in what would become the new discipline of computer science. In fact, the number of pure mathematics graduates needed in this section, along with their average age, steadily dropped as successful mechanisation took over.

Post-war secrecy and the way history was recorded means that we know far less than we would like about what the Wrens did. But they clearly held key jobs as 'registrars' – arranging the prioritisation of tasks assigned to the Colossus machines so that their use was optimised. The Wrens also operated these vast behemoths in eight-hour shifts around the clock. The conditions were awful – heat from the thermionic valves and lack of ventilation, together with extreme levels of noise from the running of tape, made for a terrible working environment, compounded by poor insulation and flooding that caused regular electric shocks.

Operators soon took to wearing rubber boots to avoid more serious harm.

It is clear that Wrens fixed and soldered the correct wiring and performed constant maintenance and engineering tasks on the machines, becoming experts at making them work and keeping them running. This was as critical to the codebreaking process as anything else in the chain, from Y Station intercepts to finished reporting.

Nor were women at Bletchley unaware of the inequity of their position. When Carmen Blacker arrived in 1942 she was told that if her weekly pay of £2 seemed low, 'it was partly due to my age, 18, and partly due to my being a woman'. The leadership constantly struggled to match the pay scales of the military, the Foreign Office and other branches of the civil service. Nigel de Grey, Bletchley's chief bureaucrat, persistently complained and tried to find ways around the system. Joan Clarke was promoted to 'linguist grade' despite her lack of foreign languages, and it was suggested that she might need to enlist in the Wrens to get further promotion and a pay rise. She declined. Hilary Brett-Smith was lucky enough to receive a man's wage on the basis that Foreign Office administrators assumed from her name that she was male.

The female workforce at Bletchley Park was also primarily responsible for the final job of the war – dismantling and destroying almost everything. The destruction of all but a few machines, which went to GCHQ, and the absolute secrecy enjoined on those who had developed and operated them, were understandable in the context of the developing Cold War. And in classically understated GCHQ style, it only took the minimum needed – two Colossus computers out of the ten operating at Bletchley – rather than the most it thought it could get away with. GCHQ and General Post Office engineers were keen to recycle the very precious contents of the machines.

This was a strategic error. Histories of computing were written in ignorance of the existence of Colossus until the late 1970s, and even then the detail of what the machines were used for was not revealed. It was widely assumed that the ENIAC (Electronic Numerical Integrator and Computer) at the University of Pennsylvania was the world's first electronic multi-purpose computer. The issue is not who got the credit – the great names in US computing, notably John von Neumann, gave Turing his plaudits but were unaware of the progress made by Max Newman and others in making his vision a reality. Much more importantly, Britain's obsessive secrecy and overemphasis on national security effectively ceded its leadership in the development of computing.

This was mirrored by the fate of the female workforce. As male labour returned from the war, women found that the technical jobs they had learnt at Bletchley were soon denied to them. We have numerous accounts from female veterans describing how they had hoped to take their wireless operating skills from the Y service to peacetime employment, but were informed that Marconi and other companies were only hiring men.

Several hundred women, mostly operators of the Hollerith card indexing machines, transferred to GCHQ. Given the numbers, Bletchley's leadership made sure they were transferred across en masse through the complex world of civil service grades. These women had operated the memory bank of

Bletchley Park, their indexes having ensured that the huge volume of disparate and piecemeal messages intercepted were an easily comprehended part of a larger picture. And it is worth remembering that judgements made about how information was stored or categorised could be much more difficult and important than codebreaking work, which was often repetitive.

But nonetheless a generation of female technicians and early computer programmers – those who ran and fixed Colossus and the bombes, as well as many of the Hollerith and Typex operators – was lost to the country's workforce. Progress in computing was taken forward, notably by Max Newman at Manchester University and at the National Physical Laboratory in Teddington on the outskirts of London, but there was no national push for innovation; the engineering progress that had already been made was confined to the narrow national security space, in stark contrast to parallel developments in the United States.

While universities in the US seized the initiative in computer science, female codebreakers in the country were regarded in much the same way as in Britain and until recently have largely gone missing from historical accounts. It was not until 2019 that Congress passed a resolution detailing Elizebeth Friedman's career, of which we will hear more, including acknowledgement that in the 1930s 'she created and managed the first codebreaking unit ever to be run by a woman'. Tellingly, the resolution stated that J. Edgar Hoover 'took credit for the achievements of Friedman and her team, leaving her work widely unrecognised until after her death'.

The picture at GCHQ was mixed. John Ferris, the organisation's authorised biographer, conducted an exhaustive study of the personnel data of post-war GCHQ and came to the broad conclusion that the agency was ahead of its peers and its time in the roles occupied by women. But it was not immune to societal norms.

When I returned to GCHQ from my discussion with a new generation of girls and a veteran of the Colossus operation, I had to conclude that the national picture seventy-five years on is only marginally better. At a time when there is a shortage of people with technical skills both in Britain and worldwide, especially in cyber security, half of the UK workforce – which was over-represented at Bletchley – is still massively under-represented today.

Today, only around 12 per cent of engineers in the UK are women. Research published by the Institution of Engineering and Technology underlines the early roots of the problem: girls at school are as interested in STEM subjects as boys until the age of fifteen or sixteen, but then something happens to convince them that these will not provide a desirable or viable career path.

As a result of this analysis, and the work of some dedicated female and male staff, GCHQ pioneered a number of initiatives to encourage women into technical careers, including girls-only competitions for cyber security and girls-only summer courses. These experiments were a great success, and it was clear to me when I visited the inaugural summer school that the absence of boys – whose tech focus at that age tends to be on gaming – made a significant difference.

There is much further to go to put right the generational problems of stereotyping in science and engineering, despite a great number of excellent attempts to do so. Honouring the largely unrecorded history of women in science is part of that process. Among its many other accolades, Bletchley Park should be feted as a great female organisation.

As GCHQ celebrated the appointment in 2023 of its first woman director, Anne Keast-Butler, the last seventy years could be seen as a gradual climb back to the heights of the Bletchley era and beyond.

11

Tolerance, Difference and Unity of Purpose

Whenever the intelligence agencies have said anything public about diversity, there is an almost inevitable reaction, albeit from a minority. The gist of this is that intelligence officers should be getting on with fighting terrorism, or whatever it is they do, instead of indulging in fashionable campaigns. But this kind of response fails to grasp two key features that have run through GCHQ's history and underpinned its success.

The first is obvious: excluding whole groups of potential employees by making them feel unwelcome inevitably shrinks the available talent pool. The second is more important: cognitive diversity comes not only from biology and neural differences but from life experience, background, education, culture and many other influences. What makes individuals effective in the secret world is a complex amalgam. Their ethnic background, sexuality, life experience, interests and allegiances are not completely irrelevant; these are part of what they bring to work. Codifying, measuring and systematising difference is a thankless task, but you know it when you see it in action in a team, and particularly when you contrast it with grey uniformity. It does not follow that Bletchley Park was a campaigning organisation for minority groups, or driven by fashion, nor that it was immune from prejudice, but a culture of tolerating diversity was baked into the codebreaking experience. The same can be said for GCHQ today.

It is a deep irony that two of the groups that Hitler was systematically trying to exterminate – Jews and homosexuals – played such prominent roles at Bletchley. At the heart of the most secret operation in British history was a gay man, Alan Turing, and a Jew of German heritage, Max Newman, with the whole operation being based in an estate built by a prominent British Jew, Sir Herbert Leon. It is enough to torment conspiracy theorists.

Confirmed bachelors

Needless to say, this secret world was neither a utopia nor insulated from wider societal and legal pressures. The distinguished mathematician Jack Good, who worked alongside Turing, later flippantly remarked, 'Fortunately the authorities did not know Turing was homosexual. Otherwise we might have lost the war.' This was a deliberate exaggeration, but it made a point.

Turing did not hide his sexuality from those who worked most closely with him, but it was nonetheless not widely known. He judged who he could trust – given that homosexual activity was still a criminal offence – notably telling Joan Clarke a few days after their engagement about his 'homosexual tendencies'. Whether the leadership of Bletchley Park knew is not recorded – it seems very likely that they did since there are quite a number of records from veterans talking about homosexual behaviour they had observed in the Park. They associated displays of affection between gay men as 'donnish' antics, the sort of thing that was commonplace at King's College, Cambridge. We have to assume that the leadership of Bletchley Park thought it irrelevant to the running of the organisation and a private matter.

Following his work on Enigma, Turing went on to do important research on the security of voice communications, including the encryption or 'scrambling' of phone calls, and after the war he was consulted by Hugh Alexander at GCHQ on sensitive issues in relation to the growing Soviet threat. All this stopped with his arrest and conviction for 'gross indecency' in 1952, along with his ability to visit the United States.

He was still producing remarkable work, and it is of course impossible to say what further contributions he might have made to GCHQ from the outside, never mind to the rest of the world. To the organisation's credit, his Bletchley Park colleagues remained loyal to him: Max and Lyn Newman were close friends, and they ensured he retained his academic position at the University of Manchester. Hugh Alexander was a character witness at his trial, describing him as a 'national asset' – a courageous thing to do under the circumstances, and delivered with the approval of Edward Travis as director of GCHQ.

But Turing's high-profile story represents a much wider loss. When we decided in 2015 to light up GCHQ's Doughnut headquarters in rainbow colours, I received a long letter from a former member of staff, 'Ian', who now lived in Germany. He had served in the RAF and joined GCHQ in 1961. After seven years of exemplary service, with glowing prospects for the future, he was interrogated on suspicion of being homosexual – one of his colleagues had reported him – then summarily dismissed and escorted out of the building. He received no support at all from anyone in authority, even his union, and no one ever followed up to check on his well-being or to show him any compassion. Not surprisingly, his health suffered and the psychological effects of this humiliation were long-lasting. While he eventually found other employment in another part of the civil service, he was surely right in believing that in career terms he never reached the potential expected of him, his prospects cruelly cut short.

It was Ian who asked that GCHQ should apologise publicly – not in some way to 'pardon' him, because, as he said, he had done nothing wrong, but to say sorry to him and to the many hundreds of others affected right up until the 1990s, when the organisation's policy was changed. He and his partner later visited GCHQ and spoke to the thriving LGBT staff group.

Throughout the Cold War there were many well-known 'confirmed bachelors' working at GCHQ; this does not minimise the injustice and prejudice involved, but illustrates that its culture was, at least on this issue, notably more liberal than the rest of government. While understandable paranoia about Soviet blackmail opportunities drove the vetting of officers in the Foreign Office, Cheltenham evaded some of the zealotry and continued to prioritise expertise over dogma.

Of course, there has also been criticism of what some have seen as the 'modern gesture' of apologising to Ian and his peers, despite this primarily being about fairness and justice for the individuals; there has never been much interest in culture wars.

But those affected also challenged and changed the organisation internally – a reminder that comfortable harmony is not always healthy and that the organisation's greatest achievements have been built on a culture of challenge and the acceptance of difference.

As importantly, this public debate had an impact on the external image of the organisation. There was a surge of visits to the recruitment website, especially from a young demographic. The point was not narrow – it was not about every gay computer scientist and engineer deciding that GCHQ was the place for them. The statements made by GCHQ – stretching back to the best features of Room 40 and Bletchley – promoted an openness of culture, a set of tolerant values, and above all the ranking of ability and merit above conformity, all of which the young target audience warmed to. Culture is a hard thing to pin down, but just as Joseph Eachus knew when he walked in the door that he would be at home in the eccentric world of US naval crypt-analysis, so recruits naturally and quickly pick up the feeling of whether they will be allowed to flourish.

For the secret world, making sure that its culture is suffi-ciently heterogeneous – or even eccentric – not only improves the chances of cognitive diversity, of approaching problems from different angles, it is essential to challenging the greatest dangers presented by groupthink.

In short, healthy diversity is the best chance of inoculating an organisation against the myriad natural forms of bias built into our analyses. The reason I first became interested in cognitive diversity was my rather slow realisation that the teams I had been involved in that were the most enjoyable and successful were also the most diverse. Working in the Northern Ireland Office on the peace process was an early introduction. The blend of local civil servants from unionist and nationalist cultural and political backgrounds, along with Whitehall secondees and military and intelligence personnel, was an unusual combination. Since both the subject and the department were unglamorous, and because many staff had a personal stake in the future of the project, people tended to stay, and there was deep expertise and shared experience, for better or worse.

That stake in the future could lead to acute pain. Coming from a community caught up in a conflict raises the chances that you will feel passionately about the issues, and increases the possibility that you, or someone you know, will be directly affected by violence. Being able to maintain emotional distance is a key requirement in such work; personal feelings have to be subordinated to the mission within a shared and clearly understood legal framework.

In any secret intelligence agency staff are unlikely to be insulated from geopolitics, whether in Afghanistan, the Middle East or any other area of conflict. GCHQ's foreign focus and its linguistic requirements mean that in general a large number of staff possess deep cultural and ethnic affinity and understanding. That is – and always was – a huge strength.

The secret world of signals intelligence has had an unusual cultural history. While in modern times it has suffered from the same lack of ethnic minority staff as other parts of government – something that both GCHQ and the US National Security Agency have worked hard to address – it has also been shaped by immigration.

'A crime without a name'

In the United States, three of the four original members of William Friedman's team who broke Japanese ciphers were first-generation Jewish immigrants. Their presence was not accidental. US government reforms of the civil service had the effect of levelling the playing field at a time when the employment options for Jewish immigrants were severely limited by institutionalised anti-Semitism. The federal government and the US military offered a way out, which these young teachers took.

In 1930s Britain the Jewish community was long established, but it grew significantly with the influx of those escaping the pogroms in Germany and countries to the east. It was socially and geographically diverse; for example, the Bogush daughters, Anita and Muriel, who respectively worked in Block A and Hut 4 at Bletchley, had been evacuated from Hackney during the Blitz, and their father chose Bletchley because he knew the only Jewish family in the village. This family hosted many religious celebrations for members of staff from the Park, including some of the famous names of Anglo-Jewry among the debutantes.

For first-generation immigrants, security vetting posed a problem, although the process appears to have been relaxed as time went on, perhaps because so much immigrant academic talent was being wasted. Max Newman's German-born father had been interned at the beginning of the First World War and he had assumed that this would bar him, but Max's value overrode any concerns.

There was also a vibrant group of younger Jewish academics who would meet on Wednesday evenings at the flat of Joe Gillis, a lecturer at the mathematics faculty of Queen's University, Belfast. There they discussed the future, personal and political. Some of Bletchley's most talented staff were regular attendees: Rolf Noskwith; Morris Hoffman, a young civil servant who had been one of the earliest members of the Federation of Zionist Youth before the war; Jack Good, a gifted mathematician and British chess champion, who was Turing's statistical assistant; Michael Cohen, a Scot who had been studying divinity at Glasgow University with a view to becoming a rabbi; and the remarkable Ettinghausen brothers, Walter and Ernest.

The Ettinghausens had been recruited from Oxford and both worked in Hut 4 in the German naval section, where Walter ran 'Z Watch'. Their team's work was critical to the protection of transatlantic convoys – Walter came to know the name of every U-boat commander – and to Royal Navy operations. Z Watch was closely involved in the hunt for the German battleship *Bismarck*, and Walter himself handled some of the doomed vessel's last messages in May 1941.

In an extraordinary juxtaposition that sums up so much of the magic of Bletchley, another German linguist in Z Watch was a softly spoken Scottish Muslim who had been the imam of Woking Mosque for some years. David Cowan was an Arabic scholar who had converted to Islam and, after the war, became a leading academic in Arabic at the School of Oriental and African Studies in London. Having studied in Cairo in the 1930s, he would certainly have been aware of the great Islamic mathematician Al-Kindi, who was applying frequency analysis to letters in codes – the same approach used by Friedman and Turing – in 9th-century Baghdad. In July 1944 Cowan translated a message by the failed 20 July plotters proclaiming that Hitler had been replaced by Field Marshal Erwin von Witzleben.

What makes the little we know of Jewish staff at Bletchley significant is not just that they represented a clear cultural and ethnic group within the Park, but because some of them were privy to secret knowledge of the attempted elimination of the Jewish people underway in Germany and the territories it occupied. It is an extreme example of a general facet of intelligence work – the holding of acutely painful knowledge, sometimes of particular personal relevance.

During the summer of 1941, Bletchley was intercepting top secret German police radio and SS Enigma messages cataloguing the systematic killing of Jews as the German army advanced into the Soviet Union, village by village, town by town. Based on these statistics of tens of thousands of deaths, Bletchley's assessment was that this was a high-level Nazi policy and that the SS and police were competing with each other as to their 'score' of victims.

The War Office believed Bletchley's view to be an exaggeration, but Churchill was reading the reports himself – where intelligence was concerned, he liked to form his own judgements, to the irritation of his senior advisers. He decided in August to risk revealing a part of the Ultra operation by speaking in some detail about 'this merciless butchery'. The scale of what was later seen as the beginning of the Final Solution was already sensed by Churchill – 'We are in the presence of a crime without a name,' he said.

In reaction to Churchill's speech, German police were instructed to send details of all future 'executions' to Berlin by hand, to avoid interception. Bletchley continued to collect material for the post-war investigation of war crimes by the Foreign Office, but the priority was to use the Ultra intelligence to win the war and bring the industrial mass-murder to an end. Having established the systematic nature of the killings, Nigel de Grey noted to Churchill that 'it is not therefore proposed to continue reporting these butcheries specially, unless requested to do so'.

By the autumn of 1941 there were enough non-secret sources pointing to the slaughter, particularly of over 33,000 Jews in the ravine of Babi Yar towards the outskirts of Kiev, that Churchill felt able to send his famous personal message to the *Jewish Chronicle* in November without risking the Ultra secret:

In the day of victory the Jew's sufferings will not be forgotten ... he will see vindicated those principles of righteousness which it was the glory of his fathers to proclaim to the world.

Two years later, on the twenty-sixth anniversary of the Balfour Declaration, he sent an almost identical message on the 'unspeakable evils' suffered by the Jewish people.

Among Jewish staff at Bletchley there was clearly some awareness of what was happening but it was partial and fragmentary, although there were reminders of what they were confronting – Morris Hoffman remembered seeing a captured German book bound with a looted Torah scroll.

Walter Ettinghausen was surprised that he and his brother passed security clearance, given their German birth. He assumed that the recruiters 'concluded, correctly, that we had an extra interest in fighting Hitler'. Writing in the 1990s he records movingly the personal impact of an intercepted signal from a German ship in the Aegean that was transporting Jews from the islands to Piraeus *'zur Endlösung'* ('to the Final Solution'). In 1944 he had never heard this phrase before, but instinctively knew what it meant. Typically of the man and the culture of Bletchley, he did not mention this to his brother or anyone else on Z Watch, until asked to write an official account in the 1990s.

The meetings at Gillis's flat were focused more on the future than on current atrocities. It was here that the Professional and Technical Aliyah Association was founded, to encourage Jewish professionals to emigrate and build a modern, democratic Israel, with Walter Ettinghausen declaring at one gathering that he would be on the first boat to Palestine after the war. He left in 1946 and, as Walter Eytan, went on to play a key role in Israeli foreign policy and public service. Gillis himself became a professor at the Weizmann Institute of Science in Tel-Aviv, founding the department of applied mathematics, while by 1948 Michael Cohen was coding messages for the Jewish Agency in Jerusalem and helped found the British kibbutz (Kibbutz HaBonim) in Upper Galilee. Many other Jewish Bletchley veterans put their skills at the service of the new state of Israel; when Noskwith saw Eytan at the UN in 1947 to offer his own expertise, Eytan responded: 'Codebreakers we have plenty of!'

Others in the group chose to stay in the UK. Many, like Newman, returned to academia; others continued to work for GCHQ, including Jack Good, Ernest Ettinghausen and Naky Doniach, who was in charge of teaching Russian during the

Cold War and whose technical dictionaries are still in use. Peter Benenson went on to found Amnesty International.

The numbers of Jewish staff at Bletchley rose significantly as United States liaison officers joined the effort. In 2016 it was formally acknowledged that the institutional descendants of the codebreakers who travelled to Israel had become important partners for the UK in the fight against terrorism and cybercrime.

The story of Jews and gay people at Bletchley and in GCHQ, or of Afghan or Muslim staff in the last twenty years, or of any staff with widely differing religious or political beliefs, says something important not only about the pivotal contribution of minorities, but about the tensions involved in serving a higher mission, particularly one that cannot be discussed openly. But above all it illustrates the power of an organisation that can tolerate, value and channel difference towards a shared objective.

12

The Power of Youth

You may or may not trust Jeff Bezos and whichever bank you choose to keep your money or your debt in. They may not trust you, and they certainly won't know you. But you are probably happy to enter into encrypted communications with them involving the transfer of your money and data across the internet. Unless cyber fraudsters get in the middle of that transaction, the communication itself is secure and unique; your neighbour's shopping transaction with Amazon or the bank is as secure as yours, but different.

The fact that you can do this – and that business across the worldwide web has boomed over the past thirty years – has a great deal to do with the work of a few young GCHQ staff. More particularly, it is another illustration of a wider approach that understands the value of youth and how best to harness young minds to challenge assumptions and pursue the seemingly impossible. A persistent theme underlying the greatest innovations of the secret world is the youthful ignorance of prior difficulty.

Of course, we take for granted the fact that we can buy from Amazon securely. But based on what we know of the history of encryption as it played out in the early 20th century, it should not be possible. After all, for several thousand years the concept was clear: for me to send you a secure, encrypted message we would both need to have the same 'key'.

At its simplest, if I sent you a box with a padlock containing something precious, we would both need identical keys, otherwise it would be of little use to you. But to ensure no one stole the key I would need to get it to you in a secure way and, obviously, get it to you separately from the box.

For encrypted messages, the key would need to be kept secret and protected from others, and available only to those permitted to read the message at either end. From the simplicity of the ancient Greeks' scytale rod device to the sophistication of the Enigma machine, this seemed an obvious and fundamental principle.

But as the volume of communications increased, it brought some obvious problems. Even for the military powers of the Second World War, the production and distribution of code-books and daily instructions on Enigma machine settings, as well as all the other encrypted systems, was a huge burden requiring a significant industry of its own.

Transition to the Cold War work of GCHQ made this problem acute. By the end of the 1950s, secure devices were proliferating across Western military organisations to guard against increasingly sophisticated Soviet intelligence and cryptanalysis. At the highest end of security, time and money could be spent distributing keys for, say, the nuclear submarine fleet. But for the ever-multiplying number of smaller devices in armies and embassies across the West, such effort and expense were becoming unsustainable. And that was an era when encryption was the preserve of governments; the vast numbers of future consumers on the yet-to-be-invented internet were scarcely imaginable.

Thinking the unthinkable

This problem fell to the part of GCHQ that had inherited responsibility for security from GC&CS, going under the equally anonymous moniker Communications-Electronics Security Group (CESG). CESG turned not to a mathematician but to one of its engineers, James Ellis, for some suggestions on better ways to distribute secret keys at scale, without apparently much expectation of anything dramatic happening. But Ellis decided to challenge the premise that it was necessary to do this at all.

If there is such a thing as a typical GCHQ cognitively diverse thinker, it was Ellis. An East London grammar-school boy, Ellis had won a place at Cambridge to read physics, despite predictions from doctors that his traumatic birth might leave him with a permanent learning disability. During the war he went not to Bletchley but to the military research laboratory at Teddington and then the Post Office Research Station at Dollis Hill, the workplace of the engineer Tommy Flowers; after the war, as Turing moved to Teddington to concentrate on computing, Ellis joined GCHQ at its first post-war base at Eastcote in Middlesex, then accompanied the organisation to Cheltenham when it moved. He gained a reputation for having extraordinary and novel ideas, most of which were entirely unworkable; his notoriously dry talks were nonetheless well attended because he was likely to drop in some challenging nugget.

Charged with the problem of distributing secret keys, Ellis began with the prevailing assumption:

> *It was obvious to everyone, including me, that no secure communication was possible without [a] secret key ... there was no incentive to look for something so clearly impossible.*

What caused Ellis to change his mind and think the unthinkable was a paper he came across from a completely different discipline, a study from the Bell Telephone Laboratories in the United States on secure voice calls (a subject to which Turing had devoted a good deal of his time during the war, with great success). The paper suggested an ingenious way of securing speech by telephone; if the receiver added a stream of noise or interference to the line and made the caller's voice inaudible, he could remove that noise later – given that he knew what it was – and be able to hear the speech of the caller.

In short, what made the difference was the active involvement of the receiver in the act of encipherment; only I know how to remove the stream of noise I have added to make your call audible to me. Ellis then went on to investigate whether this might be at least theoretically possible for ordinary encryption – could a secure message be sent without the secret exchange of secret keys?

This question actually occurred to me in bed one night, and the proof of the theoretical possibility took only a few minutes. We had an existence theorem. The unthinkable was actually possible.

Ellis had a reputation for original thinking and challenging assumptions. But he was not a mathematician or cryptanalyst, and his theory had to be tested. Eventually, Shaun Wylie, GCHQ's chief mathematician and a Bletchley veteran, pronounced, 'Unfortunately, I cannot see anything wrong with this.'

But approval of the concept by Wylie led to some years of fruitless attempts to turn theory into practice, so a fresh perspective was needed. This came almost overnight from the mind of a new recruit, Clifford Cocks, a twenty-two-year-old mathematician who had become fed up with his doctoral thesis at Cambridge and joined GCHQ in 1973, where his friend from grammar school, Malcolm Williamson, already worked.

Like all new mathematicians at GCHQ, Cocks was assigned a mentor, Nick Patterson, an Irish chess champion who eventually moved from GCHQ to the Massachusetts Institute of Technology. There he applied his codebreaking skills to biology, contributing to groundbreaking work on the Neanderthal genome.

Patterson casually mentioned Ellis's idea and its mathematical challenge. Cocks, who 'wasn't doing anything that evening', went home to his bedsit and thought about the problem, although he was not allowed to write anything down. Within around thirty or so minutes he had found a solution. This involved the production of a key by using a mathematical truism: the product of multiplying two very large prime numbers is extremely difficult to factor in reverse – that is, to work out which prime numbers had originally been used. Cocks saw that multiplying very large primes could produce the 'public' key, which could be known to anyone. Cocks then worked out and presented a formula to Patterson, by which the sender of the message could encrypt it so that only the person who knew the original prime numbers – the recipient – could decrypt it.

Cocks's school friend Malcolm Williamson refined the process further and a viable system emerged, giving life to Ellis's insight. But in practice, nothing happened because, although GCHQ shared this knowledge with its US counterpart, introducing such a radically new system at the height of the Cold War was simply too risky and too expensive.

Whether this was prudence or lack of vision is hard to say, but certainly the stakes were higher for an organisation charged with protecting life – and against the world's most sophisticated adversary – rather than simply guarding commercial secrets. A few years later US academics at Stanford and MIT developed the same ideas; in 1976 Williamson's idea re-emerged as the Diffie–Hellman key exchange, and Cocks's discovery as the RSA (Rivest–Shamir–Adleman) system in 1977. Both were patented and became the key components of public key cryptography, which underpins the security of most internet transactions. An internal debate was held as to whether GCHQ should try to block the patents or indeed patent the work of Cocks and Williamson, but this was a step too far for 1970s GCHQ. It was not until 1997 that Ellis's papers were published, and only in recent years has the scale of his achievement, and Cocks's insight in a Cheltenham bedsit fifty years ago, been recognised.

All of this in itself demanded a certain humility, a quality that runs through the story of the secret world. The experience of Ellis, Cocks and Williamson in seeing the global impact of non-secret encryption would have been galling to many, but they seem to have dealt with it with equanimity and got on with other work. Ellis died before he could be given full recognition, even internally, and his wife Brenda, their children and grand-children visited GCHQ for the unveiling of a plaque honouring his work. His wife recalled that he was 'very introspective, sitting alone thinking all the time. He lived in his own world.'

In his 1987 paper summarising his discovery, Ellis reflects on the lot of cryptographers who have to work in secret 'closed communities':

Cryptography is a most unusual science. Most professional scientists aim to be the first to publish their work, because it is through dissemination that the work realises its value. In contrast, the fullest value of cryptography is realised by mini-mising the information available to potential adversaries. Thus professional cryptographers normally work in closed communities to provide sufficient professional interaction to ensure quality while maintaining secrecy from outsiders. Revelation of these secrets is normally only sanctioned in the interests of historical accuracy after it has been demonstrated clearly that no further benefit can be obtained from continued secrecy.

Williamson left GCHQ in the 1980s to join the private sector in the United States; he worked on, among other things, new designs for digital hearing aids. Cocks remained at GCHQ until recently and is a Fellow of the Royal Society.

As for the man who had approved Ellis's idea, Shaun Wylie was a Bletchley veteran. His studies at Princeton overlapped with Turing's presence there in 1937, and in December 1940 Turing invited him to join his group in tackling German naval Enigma. Hugh Alexander, of John Lewis fame, who effectively ran the unit under Turing, remarked later, 'Except for Turing, no one made a bigger contribution to the success of Hut 8 than Wylie; he was easily the best all-rounder in the section, astonishingly quick and resourceful.' Wylie retired in 1973 and with his wife, whom he had met at Bletchley, moved to Cambridgeshire, where he taught in secondary schools and helped found the University of the Third Age. Unsurprisingly, he composed crosswords for the *Listener* magazine, using the name 'Petti' (the 15th-century Scottish word for petticoat was a *wyliecoat*).

What Ellis and his colleagues could not have foreseen in 1969 was that cryptography would move out of the secret world and become available to everyone, being of particular use to those of us shopping online. The kind of security at that time solely the preserve of nation states would be made available as a ubiquitous service for anyone with an internet connection and, today, a smartphone. The new challenge for GCHQ would be to tackle threats in this new information age from those using high-grade encryption to do harm.

As a case study, the story of Ellis, Cocks and Williamson encapsulates many of the themes of GCHQ. Ellis was given a problem without much expectation of a solution. He was an engineer, not a mathematician, but had a reputation for approaching problems from first principles, often with unorthodox results. His remarkable and solitary idea, sparked by reading a study from a different discipline altogether, was unworkable without the ideas of others. These were supplied by two very young mathematicians who had recently arrived and possessed that youthful lack of concern about what might be possible. Their collective work was not taken far, partly because of the Cold War, partly because the agency had not yet foreseen the ubiquitous technology that was coming through the internet thanks to the work of Vint Cerf, Bob Kahn and other pioneers in the American academic semi-secret world. (When I asked Cerf how he and his colleagues came to solve the problem of protocols that would enable a global network, he said, 'They asked me to solve the problem and we were too young to know it wasn't possible.')

The 'chuck and chance it' spirit

Wylie had experienced the power of youth in problem-solving while serving at Bletchley Park. As well as being three-quarters female, the workforce was also predominantly young. One of the many quotes attributed to Churchill has him addressing the staff at Bletchley and saying, 'I knew you were all mad but I never realised you were so young.' This was partly a result of conscription, but it was still remarkable. Seventeen- or eighteen-year-old women at Y stations around the country and further afield, listening to crackly intercepted radio conversations or encrypted Morse code messages, maths students straight out of secondary school, young women from department stores, industry and academia, as well as from the services: Bletchley felt youthful. And throughout the Cold War, quick in its appreciation that university degrees were not for everyone and not necessary for everything, GCHQ continued to recruit large numbers of north Gloucestershire school leavers to work on traffic analysis of Soviet signals.

Wylie regarded the greatest breakthrough in the greatest achievement at Bletchley as the work of a very young student, Bill Tutte. As the war went on, it became clear that the highest-grade German messages, including those from Hitler himself, were being encrypted not on Enigma devices but on the more sophisticated Lorenz machine. As we have already mentioned, no one had set eyes upon a Lorenz device, unlike Enigma, which had been commercially available and which Turing and others had been able to study. But in one of the greatest intellectual feats of the war, Tutte managed to reverse-engineer the Lorenz machine over a period of a few months from the unintelligible encrypted teleprinter traffic presented to him. This enabled others to construct Colossus, the computer dedicated to breaking the encoded Lorenz messages in time for D-Day, handing Allied commanders a decisive advantage. Tutte, like Cocks, simply remarked that no one had told him it couldn't be done.

The culture of youth at Bletchley was such that even Turing, at thirty-nine, seemed ancient to some, and anyone over fifty was distinctly unusual. This affected the culture and probably the informality of the place. Hugh Alexander was an enthusiastic champion of employing young people. The older and more sceptical Nigel de Grey, Bletchley's chief bureaucrat and a man born elderly, saw both sides. Youth, he said, is 'more trainable, more prepared to accept directions, better able to stand the strain, more flexible in mind – all obvious considerations'. Against this he set some of the downsides encountered in six years of war:

i) experience had a value and was none too prevalent, ii) cases of mental breakdown occurred equally among young and old, iii) both men and women are often tougher in middle age than in youth, iv) flexibility is not always as valuable as judgement.

In the context of innovative scientific and mathematical work, the fresh perspective of youth has obvious advantages. There is a reason why the Fields Medal – the Nobel Prize of mathematics – is only awarded to those under forty. The outstanding Bill Tutte's breaking of the Lorenz cipher in the 1940s and Clifford Cocks's breakthrough at GCHQ in the 1970s are reminders of how fresh thinking, untrammelled by the weight of previous attempts, solves problems. This was not only true for exceptional mathematicians; young employees at Bletchley were also encouraged to make suggestions. There was a culture of improvement, and challenge – at least on technical processes to improve systems – was encouraged. With the regular introduction of new recruits, this approach helped to guard against the homogenising, hierarchical processes that beset most organisations. One secret of Bletchley's success was to harness young talent and allow it, effectively, to lead the work.

At its best, Bletchley's way of handling this was to use young and old in parallel. De Grey reflected that creating research sections for the likes of Dilly Knox meant that they were free to harness the new, young recruits who had 'an active sense of urgency'. According to de Grey, the problem with research was that

> *it tended to cloud the practical attack, the 'chuck and chance it' spirit that might hook the fish while the more experienced fisherman still considered the colour of his fly. Both were necessary, but one should be tried not independently of, but separately from, the other.*

This is a problem familiar to most technology organisations but acute in GCHQ – how to exploit the talent of youth and give it its necessary freedom within parameters. The tendency to overanalyse, to 'admire the problem', is common within both the civil service and academic research, two of the feeder systems that help staff a signals intelligence agency. Getting the optimum balance between research and action is difficult, and it is made the more so by the fact that the understanding of technology is increasingly seen as a generational skill. For Israel's signals intelligence organisation, compulsory military service enables them to have the pick of young technical talent of student age and blend this with older, more experienced leaders, in a manner that de Grey would have endorsed.

Like anyone involved in cyber security, I have often been asked how to get a company board to understand the risks that accompany new technology. The age and background of board members are usually an impediment, for while they may be deeply versed in corporate and financial governance, they are unlikely to have a similarly profound and nuanced technical understanding of the vulnerabilities in cyberspace. Very often board members will lack the background knowledge to grasp the concepts and ask the right questions, but equally surely they can always find someone at the other end of the age range who does understand and who will challenge them. This means taking some risks that are uncomfortable. At Bletchley, it was the long-haired, dishevelled Harry Hinsley, who had joined while only halfway through his degree course, briefing slightly astonished military chiefs, who were significantly grander in the 1940s than they are now, on intelligence produced in a way they didn't understand.

Such a resolutely merit-based approach – having the right skills irrespective of age – may seem obviously advantageous. But it runs against most organisational structures, which favour longevity and experience, and find it hard to run separate capabilities in parallel. Holding the two in tension is a matter of judgement and, of course, the job of leaders – one at which the senior staff at Bletchley excelled. But out of this tension can come a great prize of youth: doing the unthinkable or seemingly impossible.

13

Thinking Like
the Enemy

If you have ever written your computer or online password on a Post-it note, or if you routinely reuse your passwords (perhaps cleverly, you think, changing a letter or adding a '?'), or if you use your pet's name or your football club's, or even if you still use 123456, then you are part of a long history of what makes codebreaking easier. And it's not your fault. The instructions you have been given do not take into account the way the enemy – cybercriminals – work, and the way users – yourselves – think.

In 2018 the recently formed National Cyber Security Centre, a new part of GCHQ, turned computer password guidance on its head. It told banks and other organisations to stop demanding ever more complex passwords from customers and then asking them to change these once a month. This might make the banks feel better, but it was actually reducing security and making life easier for criminals.

The NCSC's technical director, Ian Levy, described the existing system as daft; in effect it required the average member of the public to remember a new 600-digit number every month. Since even some of his most mathematically gifted colleagues could not do this – and he was a mathematician himself – why was it a good idea to rely on this approach for good cyber security? It inevitably pushed users to write passwords down or use very obvious words that they might just remember.

Coupled with this understanding of human nature and its limitations was an understanding of the enemy: how cyber attackers, whether criminal or from hostile nation states, thought and behaved. They routinely use computer programs to cycle through millions of possible passwords until they find the right one. The more predictable the choice of password, the quicker and easier the process.

Instead, behavioural research suggested that using three random words that meant something to the user would be harder to crack and easier to remember. Password-cracking programs are dependent on probability, storing millions of variations of likely passwords, including those that have been seen before. But three words that mean something to you personally, that have a connection in your brain but nowhere else, are less likely to be guessed. And what is the point of changing such a password regularly if it is secure (unless, of course, it has been compromised)?

The key insight is both to understand the mind of the adversary and to understand whether the security solutions actually survive contact with human behaviour. It is no use blaming users. So the creators of Enigma machines, and the armed forces who decided to use them at scale, should have thought about the operators. They should, for example, have assumed that bored soldiers would sometimes take shortcuts and not follow the instructions exactly, or repeatedly use their girlfriend's name, or express their frustration through the same choice German swear words.

A 'comedy of errors'

In fact, one of the most remarkable historical puzzles about Enigma is why the German military failed to spot what was happening on such a large scale and for so long. At the height of Bletchley's success, Allied convoys were miraculously evading detection while the German U-boat fleet was being systematically hunted down. Did it not occur to them that their communications were being read? Of course it did – there were a number of investigations in Berlin during the course of the war, most of which Bletchley Park could follow in some detail. The reasons they repeatedly came to the wrong conclusion are very instructive for any organisation facing attack or disruption.

First, there was an over-confidence in dated technology. Enigma machines were impressive when launched in 1918, and they went through a number of modifications in the two decades that followed. But the flaw that the young Scottish dancer at GC&CS, Hugh Foss, spotted very quickly – that a letter pressed on the keyboard would never be encrypted as itself – was never corrected. By contrast, British and American communications security experts – the cyber security practitioners of their day – developed a machine that learnt from the Germans' mistakes and was not broken. But in the period when the Royal Navy relied on outdated systems, these were repeatedly broken by the codebreakers of German naval intelligence, with disastrous results. Code making and breaking is an arms race, like cyber attack and defence; you have to keep up, and complacency is not an option.

Even without the Germans abandoning Enigma, some self-criticism and a few limited technical changes on their part would have made Bletchley's job almost impossible. Failure to test the machine to destruction – to think like an adversary – was only one aspect of their over-reliance on technology. The more important problem was their complete underestimation of the weakness presented by human users of these sophisticated machines.

Many of Bletchley Park's opportunities came from what Welchman called a 'comedy of errors' from Enigma operators around the world. For example, it was essential to the successful running of the bombe machines that there should be some 'cribs' – accurate guesses of pieces of text – to start with, and here military protocol and punctuality helped the codebreakers. According to Welchman:

> We developed a very friendly feeling for a German Officer who sat in the Qattara Depression [a vast desert in northern Egypt] … reporting everyday with the utmost regularity that he had nothing to report. In cases like this we would have liked to ask the British commander to leave our helper alone.

As we have seen, overestimation of human competence and failure to understand changing user behaviour continue to be key points in the cyber world where security breaks down, the fault usually lying on the designer's rather than the human operator's side. Inevitably, in any organisation as large as the Wehrmacht – the German armed forces – there were shortcuts and breaches of the rules around Enigma radio networks.

Without this mixture of boredom, laziness and misunderstanding of how the machine really kept messages secure among users, the codebreakers would probably have failed.

In addition, there were – and continue to be – cultural challenges that went to the heart of what differentiated the Bletchley and subsequent British approaches from those of their adversaries. Whenever GCHQ launched some kind of 'diversity' initiative, there would be a rumble of disapproval from some observers. This was a failure to understand either the lessons of history or the power of cognitive difference. Had the German military not been burdened by a number of deep cultural prejudices, it might at least have spotted the problems Bletchley represented and then perhaps even addressed them.

First, the Wehrmacht made little use of civilians for codebreaking, and never allowed young academics the kind of freedom and initiative encouraged at Bletchley Park. Had they turned to brilliant German mathematicians they might have realised that the huge number of possibilities that Enigma generated – giving such false statistical confidence to its creators and operators – could in fact be broken down into manageable numbers and subjected to probabilistic analysis.

The Nazi policy of racial purity also meant that immigrant scientists working in Germany such as Edward Teller and Albert Einstein had already fled, whereas Bletchley and its American counterparts were able, with some nervousness, to use the talents of even first-generation immigrants.

That Bletchley Park was also able to turn a blind eye to the homosexuality of some of its staff was deeply ironic, given Heinrich Himmler's mockery of the presence of gay personnel within British intelligence and his personal determination to send any he found within his own SS to concentration camps, where 'they should be shot while attempting to escape'.

Even without the extremes of Nazism, the culture of the German military machine, and its preferential attachment to seniority, class and loyalty over the skills relevant to codebreaking, looked old-fashioned and failed to harness the undoubted talent available in German universities.

More generally, there were aspects of the organisation of the German military and political machine that made good security difficult and spotting problems less likely. Organisationally, the Wehrmacht lacked unity, found it hard to cooperate among its separate arms and remained siloed in the wrong ways; for all the wrangles at Bletchley between the armed services and the civilians, cooperation was ultimately established on a large scale.

In Berlin there was little incentive to bring the German armed forces together, whether to do signals intelligence or to look for breaches. Had each part of the military shared all they knew from their investigations, the bigger picture would have revealed the weight of evidence – and the sheer statistical probability – that Enigma was consistently being broken, both of these challenging the deeply held belief that Enigma was secure. Instead, it was easier to explain away all possible breaches as the result of spies, an accusation that was bolstered by Bletchley's active disinformation campaign. The pieces of the puzzle were therefore never put together, largely because they were not shared and no one could see them all. By contrast, the organisational primacy of a bunch of civilians at Bletchley over the information fiefdoms of their military colleagues was ultimately game-changing.

The way in which the German military was organised dictated how they approached the problem, with betrayal being a convenient and straightforward narrative to ply. Breaches of intelligence thus came to be blamed on a series of local issues to which they could supply plausible, local explanations, all of which suited their experience of inter-service rivalry and disloyalty. The Berlin machine, which ruled by fear, paranoia and mutual suspicion, was, at the very least, unlikely to foster the cooperative innovation and creativity that we have seen to be necessary for the codebreaking enterprise.

At a leadership level, it cannot have helped that Hitler had such an inflated estimation of his own strategic genius that he considered intelligence to be largely unnecessary. His self-confidence, mirrored by many of his military commanders, contrasted with Churchill's approach. The British leader was arguably far too swayed by intelligence, but he was right in thinking that codebreaking might just turn out to be one of those unique advantages that could help stave off impending invasion and defeat. As a speculative investment decision this was not crazy, given his understanding of what had been achieved by Room 40 in the First World War. Bletchley was an asymmetric response to a huge problem, the mass encryption of data by a better-prepared and better-equipped enemy.

But leaving aside autocratic leadership, a more pervasive and contemporary problem prevented the Germans from realising what was going on. Organisations have a great capacity for collective self-delusion, one that grows in proportion to their success or power, with humility seldom their core competence. 'Someone may soon do this better than us' is an unlikely slogan for a corporate giant. But without that thought, no one can ask 'How?', let alone 'How might we stop them?'

This may partly be the hubris that accompanies power, but I suspect it has more to do with fear of failure. If you are a multi-billion-dollar tech company, you have a lot to lose; if you are a small start-up, not so much. The higher the stakes, the harder it is to think the unthinkable – that all this might collapse. Such fear of failure affects every level of the enterprise, from the founder to the new recruit. In the case of the German response to Enigma, there was a sense that it was simply too big to fail, the consequences too dire to admit and too dangerous to raise in a system that did not welcome internal criticism. Admiral Karl Dönitz, who probably came close to the realisation that his communications were fatally compromised, had too much to lose and was personally invested in the success of the Nazi regime.

By contrast, the young men and women who broke codes on an industrial scale in Buckinghamshire had the advantage of low expectation from the outside but high ambition within. They started from the premise that all networks fail and linked networks bring each other down – not a bad principle for cyber security today. But even those who believed that the job was possible were not sure how it would be achieved; they had little to lose, intellectually at least. And at its best, Bletchley constructed a system where self-disruption was made as easy as possible. This was achieved both through the healthy tension and cooperation between wildly different parts of the organisation, and through its culture. Where it mattered most, there was an openness to challenge and improvement – here was an organisation that encouraged itself to be jolted periodically – and a constant flow of new, young recruits who were taken seriously when they suggested things could be different, such was the overwhelming focus on a single collective goal to which personal pride and office politics were largely subordinated.

Perhaps, above all, Bletchley and GCHQ understood that technology derives from people and that people are the key to unlocking it. For all the brilliant mathematics, the key insight was into how the enemy's organisation was arranged and structured, who made it up, what it felt like, what they might be thinking.

From time to time, outsiders have suggested that GCHQ should give up its cyber security mission and pass it over to a new civil service organisation. But this would be a bit like having one navy for attack and another for defence. The essence of good cyber defence is understanding cyber offence: what hostile nation states and criminals are likely to do and how they think. Crudely, knowing how a hacker thinks and works ought to make you better at defence, at winning the game of cyber chess.

Learning from the enemy, putting yourself in their shoes, is therefore crucial, even if it means forgoing your favourite pet's name as a password. And that was a lesson learnt at Bletchley Park.

14

A Collider of Puzzles, Codes and Skills

Every few years GCHQ runs a weekend 'family day'. The purpose is to give each member of staff the chance to invite a small number of those closest to them to look around the Doughnut. This is important for morale – a rare moment to glimpse where mother, father, husband, wife, son or daughter works and to absorb some of the atmosphere of the place, even if what goes on in there remains secret. It is also a good way of thanking family members for putting up with the unusual secrecy that affects their lives as well as the staff member's. Not being able to visualise where someone goes to work every day, let alone to discuss that work, is an unnatural imposition.

GCHQ staff volunteer to organise the day and it involves a huge amount of effort, from security to entertainment. But it is greatly appreciated and, given that this is a visit that money cannot buy, a nice perk at minimal expense.

Among the many stalls on the internal 'street' running round the building are stands showcasing GCHQ's mission and history, and the assorted hobbies of staff. The most popular are often those featuring codes and puzzles. I remember my own children playing in the small museum, which houses a valuable collection of Enigma and even rarer machines, and trying out a scytale rod.

This simple encoding device dates back two and half thousand years to ancient Greece. Spartan leaders seem to have used it to communicate with their own military commanders, in one case ordering an errant general to return home and charging another with disobedience. A scytale rod could hardly be simpler; a cylinder or stick is cut in two and each of the pair wanting to communicate keeps half. The message writer winds a thin strip of paper – parchment or leather in the 5th century BC – tightly around the stick. A message can then be written lengthways along the cylinder. Once unwrapped, the writing makes no sense and the message can only be reassembled correctly when wrapped around another cylinder of exactly the same diameter.

At an adjacent table in the museum, visitors were trying Caesar ciphers, another popular ancient method for encoding a message, and one very familiar to anyone who tried their hand at secret codes as a child. Each letter of the alphabet is simply substituted for another, so in Julius Caesar's own system, A became D, B became E, etc. Over the centuries increasingly complex variations of this approach have been made. But even today the basics can still be applied. At the trial of Rajib Karim, a British Airways IT specialist convicted of terrorist offences in 2011, it emerged that his particular group of extremists had avoided much more complex encryption systems because they thought they were compromised by the authorities. Instead they invented their own basic Caesar cipher, which they shared within the Microsoft Excel spreadsheets that are used in most offices. The code was easily broken.

Ancient codes

My own introduction to codes and ciphers came not from maths but from ancient history at school, as trying out Caesar's cipher was a welcome distraction from translating his account of the Gallic Wars. Classical literature was full of references to attempts to send secret messages, with the writing sometimes concealed under apparently empty wax tablets or hidden in the carcass of an animal.

The Greek historian Herodotus had the most entertaining examples, and he described a very early and crude version of steganography, the concealment of a message inside a physical object or an image. As we shall see, it became a key method for espionage down the ages and is still in use in a sophisticated form by cybercriminals. Herodotus' account described the Greek tyrant Histiaeus' attempt to set up his son-in-law Aristagoras, ruler of Miletus, by getting him to revolt against Darius I, the Persian king.

Histiaeus summoned his most trusted slave, shaved his hair off and tattooed a message on his head. He then waited for the hair to grow back and sent him to Aristagoras with instructions to ask that his head be shaved. The message apparently got through and Aristagoras attempted an ill-fated revolt, which his double-dealing father-in-law offered to put down for the Persians.

Even to a schoolboy there seemed to be several obvious flaws in such methods for sending secret messages. The scytale device demanded that two bits of rod be sent to the two ends of the communications channel – not a trivial distribution challenge. Anyone guessing the method of encryption would fairly quickly be able to experiment with different widths of stick and see an intelligible message relatively quickly.

As for Histiaeus' servant, this approach looked full of holes. Having to wait until his hair grew back would rule out urgent communications and, assuming the message was tattooed, his head would be one-time use only. Male pattern baldness was quite common in the ancient world – some of the earliest attempted cures came from Greek physicians – and would restrict the availability of suitable candidates. Finally, the risks to the servant and his head, although obviously not a concern to the tyrant Histiaeus, would limit the wider adoption of this method.

But these early examples of codes reveal some general truths. First, the scytale and the Caesar cipher represent two core approaches to cryptology, and they have been among the dominant methods of code making since their invention. The first is a 'transposition' cipher, whereby the letters of the message are simply moved around or jumbled in a particular way. The Caesar cipher, by contrast, takes a 'substitution' approach, each letter in the original text being represented by another letter, following a fixed system.

While it is easy to criticise early crude attempts to create codes, they are a good reminder that perfection is rarely required in the business of making codes and securing communications. Good enough is the key; very little, if any, information needs to be kept secret forever. Even if it only takes an adversary a few hours or at most a couple of days to run through the possibilities of a particular Caesar cipher, that may be all the time needed for the message to stay secure. Having said that, the firing instructions for a nuclear missile should be a bit more carefully protected.

There are plenty of recent examples of terrorists using encryption that buys them enough time to mount an attack but that may eventually be cracked when it is no longer directly useful. This race against time was a feature of Bletchley Park, and remains relevant in modern decryption; codes in their complexity may be rewarding to study and admire, but for those applying their skills to a particular purpose they are not an end in themselves. Appreciating the elegance of a problem has to be replaced by solving it in time to do some good.

The historical devices on display at Family Day illustrate some other key problems that only came to be resolved in the last century, and in which Bletchley Park and GCHQ played key roles: how to win the arms race of encryption by processing ever-increasing numbers of possible permutations, first using electro-mechanical devices and then by the creation of programmable computers, and how to distribute the keys to a puzzle securely, quickly and on a global scale.

The origins of code making and breaking also remind us that puzzles come in many forms. The stereotype of codebreaking as a pursuit for brilliant mathematicians is an oversimplification, and a relatively recent one at that. For many centuries language was dominant and although, as we shall see, some of the approaches and even characteristics of mathematicians may overlap with linguistic problem-solving skills, they are not necessarily the same.

Further around the 'street' at the Doughnut and a colleague was demonstrating at another table how to make a siphon puzzle jug. Physical objects and pictures have been key parts of puzzling for centuries, from the famous labyrinths and mazes of antiquity to these jugs, which gained popularity over a thousand years later. Attributed to Pythagoras, they were moulded with a central column that had a concealed pipe within it, exiting at the base. The jug could be filled and used normally until it reached a certain level, at which point a siphoning effect kicked in and the jug emptied itself. A variation on these jugs became popular in medieval England, and have been marketed on and off ever since. They illustrate the affinity between 'magic', puzzles and codes, and more importantly, underline a key theme of Bletchley and GCHQ: the skills needed to solve a difficult problem are many and varied, drawing sometimes on the solitary brilliance of the individual, but more often on teamwork and the throwing together of different skills and approaches.

It is no coincidence that those involved in 20th-century code-breaking came from a wide range of academic backgrounds and had widely differing interests. It is also noteworthy that the kind of problem-solving demanded by jugs and similar physical objects is not merely intellectual but mechanical and creative, straddling the traditional academic divisions of the sciences and the humanities.

'The most mysterious manuscript in the world'

Finally, in an alcove off the 'street' at Family Day, was a small, old office desk of dark, polished wood. It is the closest thing to a shrine in GCHQ, being the desk that Brigadier John Tiltman had made for himself in Hong Kong and shipped back to England. This was not a vanity project; he wanted a desk that would suit his preference for working while standing up, which he did not to soothe a bad back but so that he could look perpendicularly down – rather than at an oblique angle, sitting – on the squared paper he used for codes.

Wounded at the Somme, Tiltman spent the rest of his life as an extraordinary cryptanalyst, playing a key role at Bletchley Park cracking Japanese naval codes and breaking into the German High Command's most protected communications system. After the war he worked as chief cryptographer at GCHQ, and from 1964 was attached to the National Security Agency in Fort Meade, Maryland, until he was eighty.

On his desk at this Family Day was a copy of the Mount Everest of codebreakers and puzzlers, the Voynich manuscript. Tiltman mentored many intelligence staff in the UK and the US, doing much to build the transatlantic relationship. Among others, he encouraged the remarkable Mary D'Imperio, a talented cryptanalyst and computer programmer at the NSA, to tackle this great challenge.

A mysterious book – some 250 pages, handwritten on vellum with numerous complex illustrations – the Voynich manuscript exemplifies many of the problems and solutions to finding hidden meanings. Acquired by Wilfrid Voynich, a rare book collector from New York, the book seemed to come from nowhere. The manuscript probably dates from the 15th century, but the secret text may go back a further 200 years. Remarkably, given its length, no cryptanalyst has really come near to explaining its contents.

The search for the meaning behind the words, if that is what they are, and the illustrations, many of which appear to be scientific diagrams, has absorbed professional and amateur sleuths for at least a century. In the context of intelligence work, the Voynich volume is interesting because it illustrates the multi-disciplinary approach and the variety of skills needed to shed light on its meaning. Mary D'Imperio set out the different avenues of attack that might unlock the manuscript: the origins and history of the document could be highly relevant to its meaning, but equally an understanding of the subject areas of the illustrations, from herbal medicine to astronomy and alchemy, could be a way in. Of course, the strange characters themselves – more than 170,000 of them – have been subjected to every kind of linguistic and statistical analysis. And every assumption needs to be challenged, including the idea that behind this is a Western language.

The effort that the greatest government codebreakers of the 20th century in the United States and the United Kingdom put in to solving the mystery of the Voynich manuscript is significant. It was not because they had a lot of time on their hands or nothing else to do, although all of them were tantalised by what Tiltman called 'the most mysterious manuscript in the world' and D'Imperio 'an Elegant Enigma'. Rather, the techniques they applied to the document – understanding its historical and geographical context, carbon-dating the vellum, analysing the frequency of symbols and numbers by computer, studying the pictures and searching for contemporary parallels – are an exercise in the breadth and complexity of puzzle-solving.

Mary D'Imperio talked about 'attacking' the manuscript because that is the language of codebreaking. Faced with an impenetrable text, perhaps an encrypted document on hostile troop deployments, attacking the problem is about much more than analysing a collection of letters, words and numbers. Looking at the context, the history and geography may help unlock the meaning, and even if decoded, a text will be useless unless the subject matter makes sense. That means not only using those who understand the language involved but also the subject matter. It is a team effort because a range of approaches, skills and knowledge are needed for a successful attack.

The riddle of the Voynich manuscript remained unsolved at the time of Mary D'Imperio's death in 2020. She, Tiltman and other giants of codebreaking remained fairly certain that the manuscript would one day be deciphered, although their background in secret intelligence compelled them to challenge every assumption, including that there was any secret meaning to it at all. They admitted the possibility that the manuscript was an elaborate hoax – a book that could not in principle be read. This is an infuriating thought that runs against all our instincts for pattern and meaning, posing the question as to how such a thing could be achieved and why? To write 35,000 'words' or groups of characters without any meaningful pattern would be a feat in its own right, and a whole new puzzle with the makings of a powerful conspiracy theory.

A journey round the Family Day stalls and tables brings you back to this central challenge for all codebreakers: how to assemble a sufficient range of skills, experiences and mindsets to break a code or puzzle, and how to work out systematically what the best methods of attack may be, while resisting the urge to make assumptions that are not based on evidence.

15

How to Dismantle a Conspiracy Theory

Does it matter who wrote the plays attributed to William Shakespeare? Even if it does, what has it got to do with the secret world of intelligence and cybercriminals?

The man who more than anyone was responsible for the creation of GCHQ's American sister organisation, the National Security Agency, thought it did matter. He in turn had been introduced to the subject by his future wife, a talented codebreaker, when they both worked on the problem during the First World War, and they returned to it many decades later during their retirement from public service. In the intervening years the couple were key figures in the secret wartime efforts to break Japanese and German codes, later turning their attention to Soviet codes, and were core to the relationship between GCHQ and the NSA. But their interest in the Shakespeare plays was not simply a pastime; it was an assertion of the science of codebreaking and a model of how to break down a conspiracy theory using the methods of science, an approach now more relevant than ever.

The story of Elizebeth and William Friedman takes us back to the eccentric world of a late 19th-century philanthropist from

Illinois. George Fabyan was heir to a textile fortune and exemplified many of the characteristics of the American super-rich towards the end of the 19th century. It was scarcely necessary for him to conduct actual business – in fact he published a book in 1905 titled *What I Know About the Future of Cotton and Domestic Goods*, which he handed out to visitors. It contained a hundred blank pages, a surprisingly self-aware comment on his good fortune and the random allocation of wealth. While he lacked formal education, he had a driven, almost manic enthusiasm for a wide range of scientific subjects, reflecting the fashionable interests of the Late Victorian age.

At the Riverbank Laboratories, which he established on his vast estate in Geneva, Illinois, he attempted to build a perpetual motion machine, initiated a programme to study wheat genetics in order to breed a variety that would grow in parched soil and set up an acoustic-testing laboratory to research levitation through sound. For all its bizarre experiments, the fact that the acoustic laboratory still survives is testament to this golden era of privately funded amateur science and what was arguably the first independent multi-disciplinary research facility. But Fabyan had one overriding obsession that linked several of these strands: the life and work of the great Elizabethan scientist and philosopher Sir Francis Bacon.

Interest in Bacon did not start with Fabyan. Bacon's direct role in the establishment of British colonies in North America, along with his extraordinary achievements as statesman, diplomat, philosopher and early scientist, inspired generations of American leaders. Thomas Jefferson considered Bacon, along with Locke and Newton, 'the three greatest men who ever lived

… as having laid the superstructures which have been raised in the physical and moral sciences'. But what really gripped Fabyan and his circle, including his friend Theodore Roosevelt, was the theory that Bacon was the author of Shakespeare's plays. From Mark Twain to Sigmund Freud, this idea was seriously entertained and discussed, developing into an ever more elaborate conspiracy theory.

The 'Baconian theory' combined the American love of Shakespeare with the emerging science of cryptology. Its proponents pointed to the cipher, or method of encrypting information, that Bacon had devised and described in his *De Augmentis Scientiarum* of 1623, and asserted that he had used this to bury hidden meanings within the texts traditionally attributed to Shakespeare.

'Knowledge is power'

Bacon's 'bilateral cipher' is truly ingenious. His fundamental insight was that each letter of the alphabet could be represented by a set of five combinations of 'a's and 'b's without using any other letters. He constructed a 'bilateral alphabet', where A = aaaaa, B = aaaab, C = aaaba, D = aaabb, and so on. If this was a simple Caesar-style substitution it would, of course, be fairly easy to crack. But Bacon realised that, since only two letters were involved, they need not be letters at all – they could be two sets of anything. Notably, they could be represented by subtly different typefaces, as long as the receiver knew what to look for.

For example, imagine that we have agreed in advance that when I type a bold letter it represents 'a', and roman letters are 'b's. If I want to send you a secret message that says 'run', I simply need a sentence that gives you the relevant three letters from Bacon's bilateral alphabet (baaaa baabb abbaa). To achieve this I can write a longer sentence using the pre-agreed division of 'a's in bold and 'b's as roman letters. Even better I can write the opposite of what I mean, for example: '**please** stay still'. When divided into groups of five 'b's and 'a's, and checked against the bilateral alphabet, it spells 'run'.

What is particularly clever is that as long as there are five times as many letters in the sentence I send you as the secret text I actually mean, I can say anything. Nor do these need to be letters at all: I could send you a picture grid where the shaded squares stand for 'a's and the empty squares 'b's, a score for a piece of music with one instrument representing 'a's and another 'b's, or in fact anything that displays two distinct representations. In a famous example of this bilateral cipher, William Friedman had a formal photograph taken of the class he trained during the First World War. Each student is either looking at the camera ('a's) or looking away from it ('b's). By spelling out the 'a's and 'b's and using the bilateral alphabet, the picture reveals 'knowledge is power', a famous saying of Bacon's. In fact, it does not quite say that because there were not enough participants and one of them was looking the wrong way – a good reminder of the human flaws in messaging systems, a recurring theme in the secret world.

In short, Bacon combined the substitution of a Caesar cipher with steganography, which as we saw in Chapter 14 can be hiding a message on or in a physical object, or in the case of modern cybercriminals, inserting malicious computer code within an image or sound file. For Bacon, the result was something that enabled the user 'to express and signify the intentions of his mind, at any distance of place, by objects which may be presented to the eye and accommodated to the ear'. This power to signify *omnia per omnia* – everything by means of everything, or anything through anything – illustrates the power of this cipher and some of the key themes of the codebreaking world.

If the key to solving Bacon's bilateral cipher, as with all codes, is to find a pattern, then the fact that this might be in letters, numbers, pictures or sounds illustrates that even in the 17th century a variety of approaches and experiences was necessary for its successful analysis. Bacon's division of everything into the two categories of 'a' and 'b' was, in effect, the invention of binary – the division into 1s and 0s – which underpins digital computing. The arrival of quantum computing, where things can simultaneously be 1 *and* 0, is therefore exercising cryptographers in the secret world and posing new problems for those protecting information.

The key problem with Bacon's system was knowing what signified the 'a's and 'b's. Although the example used above of bold and roman characters is crude and easy to spot, subtle alterations to lettering pose a greater level of difficulty. Modern cybercriminals regularly use very slight changes in URLs or website addresses to trick people into thinking they are visiting legitimate sites. For example, at a glance 'harrypotter.com' and 'harrypotter.com' look almost identical, but in the second the 'o' of Potter is a zero in another typeface and the address could therefore be registered as a completely different website. Add the possibilities offered by typefaces in other languages and subtle modulations in the use of punctuation, and each website address has thousands of potential variants that a busy user may not spot. The millions of people who have received fraudulent emails from fake government websites offering tax rebates or from lookalike bank websites asking for account credentials will be sorely familiar with this technique.

The fact that Elizabethan printing practices led to all sorts of variations in typeface in Shakespeare's First Folios therefore opened a huge potential world of hidden meanings for Baconians. They constructed an elaborate set of theories centred on the belief that Bacon was Queen Elizabeth I's son through a secret marriage to Robert Dudley, 1st Earl of Leicester, and thereby rightful heir to the throne. *Hamlet* was a semi-autobiographical account, and *Romeo and Juliet* inspired by Bacon's affair with the French queen.

One of the most dedicated and longstanding proponents of this secret alternate Elizabethan history was a doctor from Detroit called Orville Ward Owen. He became so convinced of his theories, aided by night-time 'visitations' from Bacon, that he persuaded Fabyan to finance a trip to England to excavate a site near Chepstow Castle beside the River Wye, where he believed the original manuscripts of Bacon's plays lay hidden in metal boxes. Following a second unsuccessful trip, which involved dredging the riverbed itself, he returned home disappointed, with accounts suggesting he suffered a high level of paranoia, certain that the press were against him and 'government secret agents watched him', with dark hints that the Rosicrucians and Freemasons were also involved.

Owen's story is worth considering for three reasons, beyond his dogged optimism in the face of all the evidence. First, he constructed a 'cipher wheel', whose design he attributed to instructions he had deciphered from Bacon's writings. The wheel was found in a Detroit warehouse long after Owen's death and consisted of two vast drums around which long strips of canvas were wound. On these, passages from Shakespeare and other 'disputed' authors were glued and correlated with 'key words' chosen by Owen. While subsequently shown to be completely worthless pseudo-science, it was an indication of the growing idea that analysis of code could be mechanised, with machines helping to solve human puzzles at scale and speed.

Second, Owen exemplified the rich crossover between art, religious belief and science, in the context of which so many of the advances of the next century took place. For example, his belief that Bacon had secretly discovered that Jesus walked on the water of Lake Galilee by means of high-frequency vibration led his sponsor Fabyan to commission research on levitation through acoustics. While this particular theory was quickly disproved, acoustical research at Riverbank Laboratories led to some genuine advances and established Riverbank as a leading testing agency, which it remains today.

But most importantly for the secret codebreaking world, Fabyan's support led him to ask Owen's devoted assistant, Elizabeth Gallup, to set up a Baconian cryptological research facility at Riverbank. While Gallup was another dedicated Baconian to her dying day, her cryptological research unit recruited some remarkable young women and men. When Fabyan offered the unit's services to the American government during the First World War, it became the first US facility dedicated to deciphering German and Mexican codes, with the newly arrived recruits training others in codebreaking skills and writing the manuals that would become standard textbooks for training in the secret world from the 1930s to the 1950s.

It is to Fabyan's credit that, for all his domineering style, towering ego and superficial understanding of the problems he was addressing, he was willing to recruit talent irrespective of gender and background. Elizebeth Smith, Agnes Meyer and William Friedman were to become giants of the US secret world in two global wars, then shaped some of the most secret capabilities of the 'Five Eyes' intelligence partnership in the 20th century, with Riverbank being recognised in recent years by the NSA for its contribution to 'cryptanalysis and cryptologic training during a critical time of national need on the eve of America's entry into World War I'. In its uniquely American way, funded by the private sector and driven by eccentric, grand projects, Riverbank was performing a similar job in the United States to the Admiralty's 'Room 40' at the same time in London.

William Friedman arrived at Riverbank in 1915. He was born in what is now Moldova, from where his parents emigrated to Pittsburgh in 1892, probably as part of a wider flight of educated Jews escaping a rising tide of anti-Semitic riots and pogroms. Through scholarships and government incentives to promote agriculture, Friedman ended up at Cornell University via Michigan Agricultural College. Fabyan had approached Cornell looking for a geneticist to work on his wheat project – 'a "would-be er" not an "as-is er"' – who would break new ground rather than simply repeating other people's research.

In the odd world of Riverbank Laboratories, with its weird blend of science, superstition and Victorian fairground-style freak shows, moving from one discipline to another was not difficult. Friedman seems to have traversed fairly quickly across to Mrs Gallup's Baconian cryptology unit, his main role being to photograph manuscript pages of Shakespeare Folios in the search for hidden bilateral ciphers. He was drawn both by the scientific field itself – he had been fascinated as a child by the codes within Edgar Allan Poe's poetry – and by a young English literature graduate, Elizebeth Smith, who had been recruited as part of the cryptological team and worked as Mrs Gallup's assistant.

Elizebeth – her mother could not bear the thought of her name being vulgarised to 'Eliza' and therefore substituted an unusual 'e' – was recruited from Newberry Library in Chicago. Neither she nor William had a background in cryptology or Baconian theory, but they applied themselves and became completely absorbed in the work, despite their growing scepticism about the validity of Mrs Gallup's claims and increasing suspicion of Fabyan.

William and Elizebeth married in 1917 and went on to become dominant figures in the emerging science of cryptology over the next half century, key allies of their British counterparts and towering influences in the secret world. In Elizebeth's case, her true importance has only recently been recognised.

Their path into the secret world began at Riverbank during the First World War when, as mentioned, Fabyan put his team at the disposal of government agencies in an arrangement that could only have happened in the United States. These agencies began to send through encrypted messages for analysis, with the unit making a number of practical breakthroughs. Most eye-catching was the decipherment of messages between Germany and a ring of Hindu nationalist agents in the United States that used a relatively complex transposition cipher, based on a 'key text' – which had to be in the possession of both sender and recipient – from which numbers could be generated. In this case it was a 1914 edition of a book called *Germany and the Germans from an American Point of View*, although Friedman cracked the cipher without ever knowing the work. (Self-written poems were employed by British Special Operations Executive agents in the Second World War, as their use of already-published poems was found to be too insecure.)

The trials of the Hindu nationalists, at which Friedman gave evidence, were tumultuous; in the San Francisco courtroom in April 1918 one defendant shot another and was in turn shot dead, in the words of a deputy clerk, 'over the heads of the attorneys' by a US marshal.

Riverbank was also asked to test a British War Office army cipher machine that was regarded as unbreakable and had been produced in large numbers for use in the field by the Allies. Five test messages had been generated in Washington using this 'Pletts' device, which comprised two wheels with an alphabet on each. William quickly guessed the first keyword, 'cipher', but struggled to guess the second. According to the story told by both Friedmans, William asked Elizebeth to close her eyes, clear her mind and say the first word that came to her when he said 'cipher'. She replied 'machine', the cipher was broken and the distribution of these devices abandoned.

For the Friedmans – and for us – this experience illustrated a number of key points about the minds of puzzle breakers. First, William had arrived at 'cipher' because he correctly assumed that those who had constructed the tests would make some very human errors by reaching for a keyword or phrase that was obvious. As we shall see, anticipating this human behavioural trait has become a key factor in tackling encrypted communications.

Second, in later discussion the couple ascribed to Elizebeth's word association her own creativity, and possibly her gender. While William's attempts at an answer were methodical and his guesses systematic, Elizebeth's were intuitive, visual and distanced; she thought of a small portable field device with two rotors and said 'machine'. Whatever the exact thought process – and the link to gender was tenuous – it was an important insight that Elizebeth's guess was creative, and informed by her own experience and background, an approach that was cognitively different from William's.

Letters have personalities

From 1917 Riverbank began training army codebreakers, and William Friedman produced a series of manuals, his first writing on cryptology, although Fabyan tried to pretend he was the author. These guides are still of interest to the secret world and show Friedman's early commitment to mechanising the process – a foretaste of the work of Bletchley Park. But one paper in particular that he wrote was groundbreaking: 'The Index of Coincidence and Its Applications in Cryptography'.

Although not a statistician or mathematician, William had instinctively noticed that letters have 'personalities'; not only do some letters of the alphabet appear more often than others, but they appear in certain combinations. In any given paragraph of prose, in any language, a letter in one line will appear above the same letter in the next line, apparently by coincidence. Friedman's insight was to realise that the coincidence was measurable and the frequency of letters could be plotted on a mathematical or statistical curve. This realisation was to become highly relevant to cracking codes in the Second World War on both sides of the Atlantic, and his approach is still used today.

The Friedmans became increasingly disillusioned at Riverbank, and William escaped by enlisting and going to France in 1918 to work as a cryptologist at the American General Headquarters. Lured back by Fabyan after the war, they found the eccentric millionaire increasingly erratic and disingenuous in his assurances that they could freely challenge his Baconian work. The Friedmans left in 1920, William eventually taking up a role in the War Department in Washington and establishing himself in the 1920s and 1930s as a government expert in cryptology. Elizebeth pursued a parallel track, notably breaking the codes of smugglers during Prohibition and tackling Enigma in the Second World War.

What the Friedmans had discovered through stumbling into their jobs at Riverbank was partly a lifelong love of puzzles and codes, but also a new strand of secret public service. For William, at Riverbank 'something in me found an outlet'. He went on to shape the new profession of cryptanalysis in the US intelligence world, developing the theory and applying it in practice.

But the experience with the Baconians and life with the bizarre George Fabyan also illustrated the fine line between science and pseudo-science, between secrets and mysteries, and between cryptological puzzles and pathological conspiracy. This is worth exploring because it explains some of the obsession with conspiracy theories, perhaps as prevalent now as it ever has been, and its frequent focus on – or exacerbation by – the secret world. It also points to the kind of people, or mixture of people, who are going to be helpful in solving secret puzzles and how they can set about doing so.

Which brings us back to Shakespeare, as it did the Friedmans in retirement. The fundamental flaw in the Baconian approach to ciphers was that the desire to believe the theory that he wrote Shakespeare's plays overrode any rational application of scientific method. Towards the end of their lives, the Friedmans returned to the subject of Bacon and Shakespeare, the occasion of their first meeting some forty years earlier. In 1957 they published a definitive examination of almost all the popular theories they had come across, starting with Ignatius Donnelly, a Minnesota Congressman who had done much to popularise Baconianism, along with his bestselling books on the 'real' Atlantis and Mayanism, mysteries on an epic and entertaining scale.

What the Friedmans did was to challenge these conspiracies by asserting the scientific basis of cryptology and intelligence work:

Getting a correct solution is not a matter of the cryptanalyst's thinking he has done the trick; it is not a question of opinion, but a question of proof. No solution can be taken as valid simply because the cryptanalyst says it is; he must in addition be able to show others that it is the right one. His demonstration must be unbiased, systematic and logically sound ... in short, it must be scientific.

The key test was that two independent codebreakers ought to be able to reach the same answer independently – 'Just as there is only one valid solution to a scientific or mathematical problem, so there is only one valid solution to a cryptogram.' Such a test may sound obvious, but Elizebeth Friedman wrote this from the perspective of having been a prosecution witness in rum-smuggling trials in the 1930s, where the basis of her cryptanalysis had been repeatedly questioned and misunderstood in court by lawyers who displayed misogyny and ignorance of science in equal measure.

The Friedmans' book *The Shakespearean Ciphers Examined* was a great success, winning international prizes, and remains in print. Beyond cryptology, it is a masterclass in how to dismantle conspiracy theories and assert the importance of facts. They never impugn the motives of the Baconians, some of whom, like Elizabeth Gallup, they had known well, nor do they suggest they were crazy. In fact, 'the worst that can be said of even the most bizarre of them is that they are in other respects sensible people who ... have allowed their good judgement to be undermined.' The Friedmans describe Gallup as *a sincere and honourable woman and no fraud. When we first met her more than a quarter of a century ago, we felt … that she found in her texts what she wanted to find, and by methods which might almost have been deliberately devised to assist her.*

They took seriously the pseudo-scientific claims made by the Baconians, analysing the often highly complex methods they used. In the process of disassembling the various theories they enjoyed themselves. Before the Second World War they had been asked by Theodore Roosevelt Jr, vice-president of the publisher Doubleday, to analyse a 'system' set out in a book by the economist Wallace Cunningham. The level of ambiguity in Cunningham's method was such that they were able to apply it to a passage from Julius Caesar and find this message:

Dear Reader: Theodore Roosevelt is the true author of this play but I, Bacon, stole it from him and have the credit. Friedman can prove that this is so by this cock-eyed cipher invented by Doctor C.

Their point was not that Cunningham's or anyone else's approaches were stupid (Doubleday rejected the book but Cunningham found a publisher in California). Plotting letters on a grid to try to determine patterns was a good start and is still used today. But the systems proposed by Cunningham and others allowed them so much latitude and subjective choice that their methods were worthless. Given their own extraordinary track record in breaking real codes and ciphers, the Friedmans understood that 'art' – hunch, intuition, guesswork and experience – certainly played a part, but cryptology had to be scientific if it were to be of any value. It must be based on a rigorous analysis of facts.

Their conclusion was that although Francis Bacon's bilateral cipher was indeed an ingenious and valid system, they could find no evidence that it had ever been used in Shakespeare's plays. Their final comment is that future Baconians should hand their work over for scrutiny by 'a professional who has no strong leaning to either side of the dispute'. In short, they should try consciously to iron out bias. There was no need for this warning – the Friedmans' book effectively saw off the code-based conspiracy theories about Shakespeare's work.

The Baconian conspiracy theory persisted for nearly two centuries and was, at root, driven by the belief that a person of William Shakespeare's humble background and modest education could not have produced works of such quality and breadth of learning. From a serious question about how a schoolmaster from Stratford-upon-Avon could demonstrate such wide knowledge and eloquent profundity, they then proceeded on a wild Baconian secret treasure hunt driven by bias, hunch and lazy supposition. Having selected a noble and supposedly more sophisticated candidate than the Bard in Francis Bacon, they then had to construct a reason for the secrecy, which involved a completely concocted history of the Elizabethan and Jacobean period. In constructing these elaborate theories, they ignored the most obvious conundrum; if the plays were primarily a vehicle for conveying Bacon's secrets, how is it that they have any literary merit at all?

What Elizebeth and William Friedman established in their intelligence careers and their Shakespearean hobby was a foundational approach to codebreaking. They based this puzzle-solving in science: a method must be testable and repeatable and supported by facts. While intuition, creativity and even gender might be useful and relevant to the process, they were no substitute for a robust method. Above all, the temptation to fit facts and data to preconceived patterns, assumptions and biases was the antithesis of what codebreakers and intelligence analysts should be doing. Half a century later the official report into intelligence failures in the run-up to the invasion of Iraq reached much the same conclusion.

16

Patterns and Connections

'Aren't the clouds beautiful? They look like big balls of cotton ... I could just lie here all day, and watch them drift by ... If you use your imagination, you can see lots of things in the cloud formations ... What do you think you see, Linus?'

'Well, those clouds up there look like the map of the British Honduras on the Caribbean ... That cloud up there looks a little like the profile of Thomas Eakins, the famous painter and sculptor ... And that group of clouds over there gives me the impression of the stoning of Stephen ... I can see the apostle Paul standing there to one side ...'

'Uh huh ... That's very good ... What do you see in the clouds, Charlie Brown?'

'Well, I was going to say I saw a ducky and a horsie, but I changed my mind!'

Charles M. Schulz, *The Complete Peanuts,*
Volume 5, 1959–1960

When Alan Turing and Joan Clarke lay on the grass outside the mansion house at Bletchley Park they discussed the daisies. Ever since his childhood introduction to science, Turing had been fascinated by biology, and by plant life in particular. He and Clarke had been influenced by the work of D'Arcy Wentworth Thompson, a biologist who was also a keen mathematician and interested in the way patterns were formed and changed in the natural world. The interplay of the beauty of natural, evolutionary and mathematical patterns inspired a whole generation in the study of morphogenesis – the development of pattern and shape in biology. One of Turing's last great works was 'The Chemical Basis of Morphogenesis', published in 1952, in which he sought to explain how patterns in nature, such as stripes or spirals, came about.

At Bletchley, the discussion about flowers concerned the way leaves on the stem were arranged, how they were spaced around the stem from the base up to the flower, and the flower petals themselves. Viewed as numbers, all of these corresponded to the famous sequence set out by the 12th- to 13th-century Italian mathematician Fibonacci, each being the sum of the previous two: 0, 1, 1, 2, 3, 5, 8, 13, 21 ... etc. Fibonacci numbers continued to fascinate Turing and inspire further research into the importance of mathematics as a driving force of evolution.

In the context of codebreaking, the important point is the identification of the pattern itself: order out of randomness. Buried in the unintelligible chaos of an encrypted message is a hidden pattern that reveals both the meaning of the text and, in the case of Enigma, the wiring of the machine used to conceal it. It follows that those who have a natural interest or aptitude in looking for patterns make good codebreakers. This means most of us, because puzzle-solving appears to be an abiding and deep human interest, one not exclusive to GCHQ employees.

Why puzzles?

There have been many psychological studies on why puzzles are popular. Theories range from the immediate – release of the 'pleasure hormone' oxytocin when a puzzle is successfully completed – to the evolutionary, where noticing patterns is an important defence mechanism. Hominids who failed to notice the disturbance of the leaves in the bushes were less likely to avoid the sabre-toothed tiger who had caused it, consequently running a greater risk of not passing on their genes. In earliest human times labyrinths and riddles were an important part of religious ritual: puzzling runs deep.

In the 1940s the American behavioural psychologist B. F. Skinner tried to examine this ability to recognise patterns through one of his many experiments with pigeons. He put one bird in a box and attached an automated feeding mechanism that delivered food at random and without any reference to the pigeon's actions. He observed that the pigeons kept doing whatever random activity they had been engaged in just before the food arrived. From this his conclusion was that the pigeons had projected onto what was actually a random delivery of food some particular reason or cause; repeating this, the pigeons assumed, would then cause more food to arrive.

Whatever the reason, we seem to have a deep fear of randomness. We need explanations and find the insecurity of 'accidental' causes disturbing, instinctively doubting that coincidence can be an explanation. Even the concept of truly random numbers in mathematics is difficult. This is partly because any algorithm to create randomness is not quite random, and also because the results of a random number generator may not look or 'feel' random (for example, 1, 2, 3, 4, 5 looks like a pattern but could logically be the result of random generation).

Our thirst for patterns that make sense of the world around us is, however, not without its difficulties. Many years ago, before smartphones and social media, I had a civil service colleague who subscribed to a quarterly magazine about 'mysteries'. In practice these were long-winded conspiracy theories on an impressive array of subjects from the familiar – moon landings that never happened or the sudden proliferation of crop circles – to the niche, such as the disappearance of the Irish racehorse Shergar in 1983.

I don't think my colleague, an intelligent and very balanced person, necessarily believed all or even most of what he read. His main motivation lay in the satisfaction of seeing a mystery resolved in a complex and apparently coherent way. This was a form of puzzle, making sense of real events that are hard or even impossible to explain. He derived the same pleasure from this that others get from crime fiction or true crime podcasts, the difference being that his publication projected a fiendishly clever plot onto a real event.

If these theories reinforced his general suspicions of authority and government, then so much the better. Growing up in the nationalist community in Northern Ireland, he had particular reason to question the truthfulness of the state and to assume that its agencies were capable of conspiracies directed against his part of the population. The fact that some of these state agencies were secret simply fuelled his interest in conspiracy and made the theories impossible to disprove. So his background and life experience inevitably shaped his approach to problems or puzzles.

Like the pigeons in Skinner's experiment, we may easily see patterns where there are none. We are all prone to some degree to 'apophenia' – the tendency to impose patterns and connections that lack a basis in reality. Gamblers do it instinctively, hence their lucky item of clothing or their superstitious pre-game ritual. If they followed the strict rules of probability they would never gamble (probably), and there would be no fun in it. Perhaps an excess of caution was also sensible where sabre-toothed tigers were concerned.

Our tendency to see patterns around us moves on a spectrum from Turing and Clarke staring at a daisy and seeing genuine mathematical and physical evolutionary patterns in nature, through Linus and Charlie Brown's observations on clouds, to Hamlet's attempts to persuade Polonius he is mad by describing different animal shapes in the sky. Hamlet, or Shakespeare (or even Bacon if you believe he wrote the plays), was on to something. Pareidolia, a subset of apophenia that tends to relate to hidden messages, is at its extreme a psychosis and frequently the expression of paranoia. The famous Rorschach inkblot tests are in practice an attempt to harness our innate tendency to pareidolia, our desire to impose meaning, the basic hypothesis being that the meaning or pattern we describe when shown a random inkblot says more about us than the inkblot.

Conspiracies and mysteries

Seeing links and patterns and conspiracies where there are none is an occupational hazard of codebreakers and the intelligence community in general. It is the job of agencies to uncover bad things that are deliberately hidden by those doing them, so by definition they look for plots and conspiracies. But an intelligence analyst who approaches any problem with a preconceived expectation of a great and complex conspiracy is likely to be misled. Trying to keep an open mind and seeing things at face value is a core skill of anyone building an investigation from fragments of information.

There is a dissonance here between the intelligence world and the rest of society. Viewed from the outside, the secret world is fertile ground for pattern conspiracy theories. Secrecy and state power are a potent combination, which is why they need to be strictly limited and overseen in a free society. But they are also fascinating and entertaining in the way that all great puzzles are, which is why journalists are natural conspiracy theorists. Complex conspiracies make much better, more compelling stories than the alternative explanations, which are usually a mixture of mundane bureaucratic incompetence, systemic failure and bad luck.

Viewed from the inside of a large public organisation, things look different. Of course, it is possible to mount a complex operation in secret, Bletchley Park being an unusual and largely unrepeatable wartime example. But in general, the kind of conspiracy theories that gain currency proceed from premises that are absurd and, even if these are accepted, the conspiracies themselves would be unimaginably difficult to mount.

One of the most popular conspiracy theories during the recent pandemic, that 5G mobile networks are the cause of Covid-19, is a good example. The premises that the theory is based on – either 5G radio waves suppress the immune system, allowing the virus to take hold, or somehow the virus itself is actually transmitted across radio waves – are so far removed from all that is known about science and the natural world that they are remarkably hard to discuss. The radio waves involved in 5G are at the low-frequency end of the electromagnetic spectrum and have never been seen to damage cells in this way. And unless everything we know about physics and biology is wrong, it is hard to see how a virus could travel by radio wave.

But even if the premise that is believed as an article of faith is knocked down by a rigorous testing of facts in the style of the Friedmans, the conspiracy would require mass participation by governments and telecoms companies on a scale quite unmanageable and impossible to keep secret. Given the scale of the pandemic, it would suggest a level of excellence in delivery that no large organisation, let alone whole governments, have ever actually achieved. Most conspiracy theories are organisationally impossible, never mind implausible for other reasons.

Significantly, one of the leading proponents of the 5G Covid theory in the UK said in a long YouTube presentation that 'God has blessed me with the ability to bring disparate pieces of information together that puts the puzzle together and makes sense of it.' Outside of the cynically or politically manipulated conspiracy theories – of which there are many – this sense of a puzzle being solved is a powerful incentive to see patterns where none exist.

It is not a coincidence that the pandemic has spawned so many conspiracy theories. Following excessively traumatic events, we would like to make sense of the shock and seek comfort in explanations that imbue them with greater purpose. After 9/11, the death toll and the brutality of the attack on the American way of life were so jarring that conspiracy theories proliferated. To be told by the US government's 9/11 Commission that there were a number of fanatical individuals from a group called Al-Qaeda who hated America enough to carry out such an attack, that the US military and intelligence apparatus had not really joined the dots of the information they possessed nor imagined that such a thing could happen, seemed inadequate and underwhelming. The explanation did not live up to the extreme trauma of the event.

To some extent the intelligence community has been the victim of its own popular portrayals of conspiracy. James Bond films would not be worth watching if Bond were not unfailingly successful in tackling something highly complex. Two hours of watching him doing a morning's unrewarding desk work followed by a minor breakthrough or setback in the accounts department would hardly be exciting viewing, nor would following 007 as he books a parking space at the office. Everything has to work seamlessly in a highly complex plot, albeit with the occasional glitch for dramatic suspense.

But government organisations are in fact beset by trivial problems. My wife, who has spent a long time observing the intelligence community at one remove, watched *The Bourne Ultimatum* one Christmas. She said she could suspend belief in all the extraordinary action sequences and labyrinthine plots and sub-plots until the final scene. In this, Pamela Landy, a high-minded CIA officer, is about to fax a pile of incriminating documents about extreme corruption by another officer. She locks the door of an empty office as he and his colleagues raid the building in an attempt to kill her, and with explosions all around, the papers transmit just in time. Anyone who remembers fax machines will know how unrealistic this is; the chances of one end misdialling, being engaged or out of paper would have been high enough to undercut the entire storyline of the film.

From an organisational perspective, the more elaborate the conspiracy, the less likely it is to be credible, so the simpler explanation is almost always to be preferred. Governments struggle to coordinate complex processes, as the pandemic repeatedly illustrated. The desire to see a conspiracy and the desire to believe something are traits that intelligence analysts are supposed to guard against.

But in the case of encrypted communications, there has to be a single truth buried somewhere in the code if it is to be useful to the sender and recipient: there must be a 'right answer'. A coded military command is not much use if the message can be interpreted as either 'advance' or 'retreat'. Truth actually matters.

For codebreakers, the existence of a code implies conspiracy; someone has thought about a way to hide something. Complexity is a given for the codemaker because in principle, the greater the number of possibilities, the harder it will be to crack.

The key to solving this complexity is to break it down and get inside the mind of the opponent, to think like the puzzle-setter. Once a set of puzzling words is written down, perhaps in a squared grid, it begins to be possible to create some order and to think yourself out of the 'normal' way of looking at them. Suddenly, words are seen as letters; it may be possible to see repetitions or connections, or possibly the letters may be seen better if they are represented by numbers.

As the GCHQ puzzle-setters cheerfully volunteered in their hints for their puzzle-book readers, knowing something about their own interests, ages and backgrounds is also relevant. Most puzzles will involve sets of things or a series of letters or numbers. Finding the connections or exceptions will be easier if the number of possibilities can be reduced. So for GCHQ puzzlers, the fact that they like particular authors (as we have seen, Tolkien, J. K. Rowling, Shakespeare), or particular films or TV shows (from Bond to *Doctor Who* to *Star Wars*), may help. Their life experience suggests they will refer to sets such as the Periodic Table or Morse code or US presidents, and game references may be to *Scrabble* or *Monopoly*. Of course, it would be a mistake to overdo the stereotypes, partly because there will be plenty of exceptions in a genuinely diverse workforce, but also because a shrewd puzzle-setter may decide to step outside these expected frames of reference. But that itself may show up as an interesting break in the general pattern.

The same principle that applied at Bletchley Park and in harmless puzzling applies in the harder end of intelligence work.

Understanding how an eighteen-year-old radicalised Islamist extremist thinks, and what experiences have shaped him or her, may be key to understanding how he or she communicates. As the 9/11 and Chilcot inquiries showed, the failure of Western intelligence agencies has often had a great deal to do with cognitive dissonance – an inability to step out of a mostly white, middle-class, university-educated background and into your adversary's shoes, no easy task when this adversary may be both culturally different and also deeply paranoid. It therefore helps that puzzle-solving and pattern recognition may be on a spectrum that covers both the extremist and the cryptanalyst. Ultimately they are involved in the same cognitive game.

17

Only Connect?

'You have to have a fully open mind,
like a child or a philosopher'

Victoria Coren Mitchell

Imagine you are looking at some intercepted communication that you believe is likely to be about a planned bombing. The bits of it you can read make little sense and the only word that keeps cropping up is 'battery'. Anyone with a layman's knowledge of bomb making will naturally think that this must be a reference to the bomb's battery detonation. But there are many other possibilities, from battery as a group of things and battery as physical harm, to the various slang uses of the word in different countries. Maybe it is simply code for something else; Al-Qaeda terrorists regularly used the word 'wedding' as a coded reference to a planned bombing. Or perhaps it stands for a place and a possible target, maybe Battery Point in Australia or Battery Park, a popular tourist destination in Lower Manhattan not far from the World Trade Center.

The analytical skills required to assess an intercepted text, even when it has been decoded, demand a mixture of lateral thinking and as wide a range of knowledge as possible. Resisting the temptation to leap to the obvious conclusion and keeping an open mind while cycling through possibilities take discipline; and Battery Park is likely to spring to mind if you are more familiar with Hobart, Tasmania and New York than Gloucestershire. Experience and background matter, as well as cognitive approach. Some of this can be automated, but current versions of AI are not good at lateral thinking. They will tell you what is most probable and most obvious, but that is precisely what you may not need.

'The back attic of the brain'

The thought processes involved in making these links will be readily understood by viewers of *Only Connect*, the fiendishly hard and incredibly popular TV quiz show. The host, Victoria Coren Mitchell, defines the lateral thinking required to succeed in a way that very closely matches the approach of good intelligence analysts and codebreakers:

> *[the ability to find connections] in the back attic of the brain; word associations; distant memories chiming; putting together the knowledge of one clue with a vague sense about another.*

The way to achieve this is not necessarily frenetic activity, although that may seem the case in the TV studio, where teams are working against the clock. In fact, the key requirement is the one Elizebeth Friedman displayed back in Riverbank, Illinois, in guessing the second word in a sequence: emptying her mind and allowing the connection to emerge unimpeded. As Coren Mitchell puts it, to be successful 'you have to have a fully open mind, like a child or a philosopher'.

Child-like approaches to the world and play in general are critical to innovative thought and creative problem-solving. I was once showing a political visitor out of the Doughnut and he stopped to look at the immensely intricate – and stunningly large – model of the building, designed and constructed by the Lego club. He had already been exposed to a 'capture the flag' hacking exercise, and he gently suggested that we spent all of our time playing. He missed the point that runs through the stories in this book, that puzzles and play do not constitute a world separate from serious work – they form a continuum, and one enables the other.

Games at Bletchley were an integral part of the work. There were palindrome competitions, which the young mathematician Peter Hilton won with 'Doc, note: I dissent. A fast never prevents a fatness. I diet on cod.' It is no coincidence that Hilton did not use pencil and paper, but visualised this over many hours with his eyes closed, a technique he also applied to codebreaking. As Coren Mitchell points out, the brain can be trained to some extent, and practice improves agility.

Knowledge also plays a part, and a diversity of the stuff counts for much; as the GCHQ puzzle-setters readily concede, understanding what they know about and are interested in helps narrow down the possible answers to the tests they set. By contrast, if the question is about the latest drill music popular with teenagers, a bunch of middle-aged men may struggle. So a breadth of ages, upbringings and experiences will inevitably help, as any good pub quiz team knows. And clearly, understanding the culture, background and mindset of a terrorist will help in the understanding of his or her communications.

But *Only Connect* also challenges teams to work out the category or nature of the connection – factual, historical, imaginary and so on – and then find the answer. In other words, to work out what the question is and then get to the answer. This has clear parallels in the intelligence analysis of data, where establishing what kind of information is being scrutinised is the first task, well before trying to work out what it actually means. In this, the different ways of looking at the world that we have already discussed can be hugely helpful. To some extent these can be taught, or at least developed through practice, and the popularity of puzzling suggests that many people have this aptitude.

There may of course be limits and, as Coren Mitchell says, 'Some things can be beyond the reach of one's brain.' In her case, she notes that she cannot picture which room is above another in a house: 'Geographical layouts won't construct themselves abstractly in my mind.' Self-evidently, her facility with lateral thinking might be very useful in the intelligence process of finding a hostage, for example, but she would not be the best person to send into the building to carry out the rescue. It could be a drawn-out process. And this leads to the most obvious feature of *Only Connect*, which also applies to signals intelligence work, namely the importance of the team.

Contrary to the popular portrayal of spies from Bond to Bourne as solitary heroes, technical intelligence work is the ultimate team sport. The image of Alan Turing as an isolated genius in *The Imitation Game* is Hollywood's version of history. Turing was certainly a genius, and alone in many ways, but from the Polish mathematicians whom he visited in Paris early in 1940, through the academic colleagues around him at Bletchley, to the wider cast of engineers and industrialists who put the ideas into action, Turing was very much a team player. It is difficult to say to whom any particular cryptographic breakthrough belonged – a concept described by Tony Comer, the former GCHQ historian, as 'the Enigma Relay'. In short, the cracking of Enigma, like so many achievements in the secret world, was a relay race rather than a glorious, egotistical sprint.

We have already seen that a mixture of cognitive approaches, knowledge bases and experiences helps in solving puzzles. But the dynamics of putting the individuals who possess these qualities together for maximum impact can be challenging. Coren Mitchell observes:

> *I've lost count of the teams where somebody on the captain's left or right – someone brilliant but not pushy – keeps muttering the right answer but the captain doesn't listen and keeps taking more clues. That makes very entertaining television but is not optimum strategy.*

I observed similar behaviour in a range of government organisations, where deeply technical and expert talent often refused to put itself forward. I learnt over the years that in almost any meeting in signals intelligence, there seemed to be 'someone in the corner' reluctant to say anything. When I asked a detailed question about how something actually worked, as opposed to its purpose or effect, all eyes would turn to this person, who was just as likely to have made the key contribution as the person at the front presenting the success. The ability to value technical talent that does not self-present in the way thrusting organisations demand is a challenge.

If it is possible to generalise at all about the staff of signals intelligence organisations, it is certainly true to say that on a Myers–Briggs Type Indicator test there would be a far greater representation of introversion than in the vast majority of other organisations. Whatever one thinks about the definition of personality types, there is a clear overlap between some neuro-diverse conditions, the generally reflective work of codebreaking and the characteristics of Jungian introversion – the gaining of mental energy through internal reflection rather than external social interaction.

The story told by Mark Rylance's character in *The Undeclared War* – Peter Kosminsky's TV drama set in GCHQ – about his proposal to his wife is not fiction. It is true that a shy linguist in the Cold War proposed to his future wife by sending a note via the pneumatic Lamson tube system to her desk in the Soviet air force analysts' office. Of course, shyness is not exclusive to GCHQ, but it is somehow less culturally surprising, and the ingenious solution particularly appropriate.

The introverted organisation

This organisational characteristic has organisational implications. Much has been written about personality types within large enterprises. Most private and public sector organisations tend to favour extrovert behaviour in their leadership and management. Articulate, confident individuals promote themselves. Very few organisations that I have come across outside GCHQ and the NSA contain such a preponderance of introverts. And introverts flourish in smaller social interactions where they do not need to – and do not like to – take the stage. One of the reasons why Bletchley Park's 'Huts' functioned so well in knowledge sharing was that much of the time interaction took place at a personal level. Large formal meetings, in which extroverts thrive but introverts tend to wither and disappear, were kept to a minimum. This was not an official decision, but one that simply happened, partly for security reasons but also thanks to a way of working that had developed organically or grown out of the academic world of Oxbridge and other universities.

The insight that introverts have something to offer has become more fashionable in recent years, although it is rarely acted upon in the corporate, administrative or political worlds. Most recruitment processes prize articulacy, self-confidence and traditional approaches to thinking over cognitive difference and introversion. This is partly because cognitive difference may be accompanied by some disruptive downsides that are hard to manage. It may also be because the non-neurodiverse feel threatened by something they struggle to understand themselves, compounded by the natural bias of extroverts to assume that people who say little have little to say.

Even where they actively try to recruit different types of thinkers, large organisations tend consciously or unconsciously to homogenise and school new talent to fit their prevailing culture. It takes a particular and conscious effort to resist this natural bias. I have visited a number of tech giants over the past few years and have always been impressed by the high quality of their mostly young staff; but I am also struck by a sense of cultural uniformity. These organisations do not breathe internal difference and dissonance. It is at the very least oddly ironic that diversity flourished in an organisation overseen by the civil service machine, albeit one that, in 1939, was given the funds and freedom to recruit as widely as it pleased.

One of the keys to GCHQ's success, inherited from Bletchley and its predecessors, has been not just the tolerance but the celebration of such different approaches and the harnessing of different personality types within complex teams. It would be wrong to suggest that GCHQ actively sought out neurodiversity in the 1940s, fifty years before the term was coined. There was no system or programme either to recruit or support staff simply because they were diverse in cognitive approach. Rather there was an instinctive and growing recognition of the correlation between what was often classed as eccentricity or difficult behaviour, and the success of Room 40 in the First World War, Bletchley Park in the Second, and GCHQ through the Cold War and into the internet era. Much of that insight started with Alastair Denniston, GC&CS's first director and a veteran of Room 40, who was himself a somewhat diffident and self-effacing character. No doubt circumstances also played a part; so many obvious and more conventional candidates had been called up to the Western Front that the Admiralty had little choice but to dig deeper, and in stranger corners, than they might have wished in searching for talent.

Denniston not only recognised the shift from languages and words to numbers and machines, he understood the value of this mixture of minds and approaches – and its superiority over the organised military hierarchies in tackling the immense task at Bletchley – and it was he who enabled it and defended it against regular criticism. Seeing off the anger of his deputy at some bad behaviour by a senior and elderly codebreaker, he commented:

> *After twenty years' experience in GC&CS, I think I may say to you that one does not expect to find the rigid discipline of a battleship among the collection of somewhat unusual civilians who form GC&CS. To endeavour to impose it would be a mistake in my mind and would not assist our war effort, we must take them as they are and try to get the best out of them. They do very stupid things, as in the present case, but they are producing what the authorities require.*

Tolerance of difference, with all its difficulties and frustrations, was key to his leadership. He frequently infuriated those around him, and eventually he was dislodged, unsuited to the industrial scale of codebreaking that emerged at Bletchley.

But his ethos of tolerance continued in parts of post-war GCHQ and shaped the reputation of an organisation where 'nerds' and eccentrics might not only do useful work but also feel at home. Secrecy helped in this – it was perhaps easier to tolerate and encourage unconventional approaches and, even more so, unconventional behaviours in an organisation that was not open to public scrutiny.

More recently, and reflecting the increase in research into neural difference and autistic spectrum disorders over the past few decades, GCHQ has actively sought to recruit this kind of diverse workforce and to support it in a structured way.

But one other factor made the use of teams in the secret intelligence world unusual. The selection of staff at Bletchley Park was slightly random and was driven by availability in the context of chronic national labour shortages, with no one expecting the enterprise to be long-term. Either Britain would be invaded – a likely prospect in the very early stages of the war – and critical staff would be evacuated; or the Axis powers would be defeated, after which the place would be closed down. A sense of precariousness and short-term survival is a useful corporate quality, but it also compelled the organisation to make the best of what it had been given.

Nor was it entirely clear precisely what jobs were required, still less what skills might best fit them. Creating new types of work is, of course, familiar in GCHQ and a popular theme in the modern tech world of artificial intelligence. But one thing marked out Bletchley's experience: the intense secrecy meant that managers could not easily shed staff. It was simply too dangerous to risk the secret of what was going on being spread, so personnel – even those who were miserable – found it very difficult to leave. The result was some unhappiness, but a great deal of effort spent in getting the best from those who had been recruited to the organisation, for better or worse. There are parallels for GCHQ and the wider civil service, not because people cannot leave but because firing staff in the public sector is quite unusual.

The pool from which Bletchley was selecting personnel was unpredictable, meaning that the teams had to do all that they could with the 'somewhat unusual collection of civilians' that Denniston found or assembled, with quick hiring and firing not an option. Yet this somewhat ad hoc approach, which goes against the grain of received modern corporate wisdom, worked – sometimes just about, at other times spectacularly well.

How these teams in the secret world worked together, how skills were blended, why it worked and still works, constitute the unwritten history of Bletchley Park and GCHQ. It is clear that many of those involved saw what they were doing as both deadly serious and a shared intellectual game or puzzle. Alan Turing, Jack Good, Dilly Knox and others explicitly refer to the element of playing a game, albeit a monumental challenge with high stakes. But the thought processes and complementary skills were no different from those required for more recreational pursuits.

So it is a nice irony that *Only Connect* is a good exemplar of these approaches to intelligence problem-solving as a team. And a few moments emptying the mind takes you to some remarkable connections to the world of codebreaking.

The title of the TV show comes from the novel *Howard's End*, published just before the First World War, in which E. M. Forster explores dislocation and dissonance: 'Only connect the prose and the passion and both will be exalted ... Live in fragments no longer. Only connect ...'. Forster's persistent fears that technology was leading to dislocation and rapidly accelerating the disappearance of a kinder world would have appealed to the creators of intelligent machines at Bletchley Park. For some of his life Turing moved in the same world as Forster at King's College, Cambridge, and Forster's novels were among those found on his bookshelf following his suicide. As the mathematician Andrew Hodges wrote in his acclaimed biography of Turing, both men were profoundly preoccupied with the same connections: 'the mind and the body, thought and action, intelligence and operations, science and society, the individual and history'.

It is entirely appropriate that the nerdishness and trivia of puzzling sit alongside deeper human connections and more serious work; in the secret world they are never separate.

Coren Mitchell cannot remember why the show is called *Only Connect* and doubts whether much thought was given to it at the time. But, she says, 'I do like the sense that linking a bunch of clues about weasels and 1970s disco might somehow serve a higher human purpose.'

18

Systematising Brains

*'Sometimes it is the people no one can imagine
anything of who do the things no one can imagine.'*

Words spoken by Alan Turing in *The Imitation Game*,
and now widely quoted, but never said by him

GCHQ's iconic doughnut-shaped headquarters – most recently imitated by Apple for its headquarters at Cupertino – is surrounded by a large car park, reflecting the vagaries of public transport in Gloucestershire and the flexible working hours of staff. Once the building was opened the car park was always over-subscribed, and lack of space usually ranked high on the list of staff grumbles. Until Covid showed the secret world how much work could be done from home, an ingenious and, unsurprisingly, highly complex online booking system tried to balance operational need, fair access, environmental concerns and security. It is a work of art in itself.

For one of my colleagues, looking at the car park presented a daily challenge: 'I need to organise it in my head and make sense of the colours and shapes and categories of the vehicles.' We were at the time discussing autism and neurodiverse conditions from a window overlooking the perimeter. He and other neuro-diverse staff taught me a great deal about the advantages of seeing the world differently and the power of compulsively pattern-seeking brains.

His natural instinct was that the parking of the cars was not random and could be systematised. He was right. Not only was a computer program using human-constructed algorithms to allocate spaces, but the choice and positioning of cars was dictated by all sorts of hidden trends, including income, taste, fashion, friendship and daily routine. The algorithms of econom-ics, human behaviour and other external factors, as well as contemporary car sales, reduced the genuine randomness to a minimum. Most people would not even think about this, or if they did they would not expend the mental effort of trying to see patterns in the car park. But he felt a compulsion to do so.

By contrast, the press office would regularly get calls from excited journalists saying 'they had heard rumours' that the GCHQ car park was unusually full – was there a crisis happen-ing? The question was always misplaced, as they were being led by ill-informed urban myth rather than analysing data sources and coolly understanding the complexity of the connections.

The benefits of the 'hyper-systemising' instinct are obvious in any job that involves assimilating large amounts of disparate information. Intelligence operations are often compared to jigsaw puzzles. Agencies very rarely possess anything like a complete picture and they are more likely to begin with a snip-pet of information – a single piece of data about a suspected terrorist overseas, perhaps part of a phone number or the make of a tablet device. From this fragment, the job of an analyst is to build a picture, establish patterns and find connections.

The galaxy of neurodivergent skills that I encountered when reading about the idiosyncratic staff of Bletchley Park or my daily dealings with their modern-day counterparts was fascinating in itself as an insight into the human brain. Both sets of people shared common features, particularly the impulse to create systems and patterns, and their unusually intense focus.

Before I arrived at GCHQ I had heard of dyspraxia but not met anyone who had been diagnosed as dyspraxic. Dyspraxia affects the communication between the brain and the body's motor functions but does not in itself alter intellectual ability. It can lead to clumsiness and lack of physical coordination as well as extreme difficulties in organisation or following instructions in the correct order.

Dyspraxia was best illustrated to me by a former colleague, who had recently retired from a lifelong career as a highly talented linguist. GCHQ is the largest recruiter of linguists in the country and covers an extraordinary range of rare languages. Like many of her colleagues, she enjoyed the process and challenge of learning a new language, and had lost count of how many she had picked up during her career.

But she explained that she struggled with basic organisation:

My cooking is a disaster unless I follow a well-tried and preferably unvarying written recipe. Cookery programmes are useless to me – I panic. My short-term memory is poor. But my long-term memory is very, very good, and that's a huge

*blessing for a linguist. I can remember large vocabularies in
different languages.*

Another colleague would become visibly stressed if I asked
him to produce a broad overview of a problem, as he found
structuring it painful. But I learnt that if I asked for a detailed
and deep dive into one particular strand, he would come back
with something absolutely exceptional. It was the organisational
or strategic framework that eluded him; his particular systematising could only take place within a very narrow spectrum.

It is less surprising that dyslexia and dyscalculia, its mathematical near equivalent, are prevalent in an organisation like
GCHQ. A significant number of apprentices joining straight
from school had the classic signs of dyslexia and told the same
story: they had performed badly, found written work or reading
difficult and fell behind. This led to demotivation and sometimes
behavioural difficulties. The attraction of an apprenticeship was
that they could get out of a system they found stifling, and do
something active and practical.

The symptoms of dyslexia are now well known – the jumbling
of letters and numbers, and difficulties in making syllables and
words out of apparently random and unconnected letters. But
the upsides are also becoming better understood and are highly
relevant to intelligence work. Because the world of letters is
either closed to dyslexics or a titanic struggle, dyslexic brains
tend to compensate by mastering other areas. Liberated from
the constraints of linear text, some of the dyslexics I have
worked with can often recall greater detail and are more likely
to see a problem holistically, distilling what matters most. They

also tend to be better at visual pattern recognition than their non-dyslexic colleagues, better at seeing the 'odd one out', are sometimes more reflective and creative, and are certainly ahead in spatial awareness and spatial imagination.

In an intelligence world, where a huge volume of information is thrown at you, these are invaluable skills and techniques to have in any team. For example, having a good spatial imagination is important – being able to see a map and instinctively grasp how far apart things are, whether X could have easily visited Y, is an obvious advantage.

At the more well-known end of the autistic spectrum are those with varying degrees of Asperger's syndrome. Not surprisingly, some of my colleagues had studied the condition in great detail and were knowledgeable guides to the voluminous research into autistic spectrum disorders (ASD) carried out in recent decades. Much of that research is controversial, but some common themes emerge about the beneficial aspects of ASD that ring true for staff in the secret worlds of Bletchley and GCHQ.

At its extreme end, a computer scientist introduced me to studies of visual acuity among those with Asperger's, some of which suggested that they were more likely to be 'eagle eyed'. The researchers meant this literally. A controlled study suggested that participants with Asperger's possessed an ability to see almost comparable to birds of prey. If the human norm is to be able to see a particular object clearly and in sharp detail at five feet (20:20 vision), birds of prey typically see the same objects clearly twenty feet away (20:5 vision) and with heightened awareness of colour. Research carried out at Cambridge more than ten years ago put the visual acuity of some people with Asperger's at around 20:7.

My interlocutor thought that he and some of his friends with Asperger's might possibly have slightly better long-distance sight than other contemporaries, but he did not see this as a significant skill or of much use in a job where he sat at a computer most days. For him, what mattered was intense, some-times obsessive, attention to detail. Alongside that came a desire to create patterned order – to make sense of chaos and the sensory overload coming at him.

These very specific skills are reportedly employed by the Israeli armed forces in a military intelligence group – Unit 9900 – that recruits autistic national service men and women. Their ability to decipher complex and blurred satellite imagery, for example, is well ahead of their contemporaries and any avail-able software. Relentless focus on the detail of what can or cannot actually be seen, resisting our normal tendency to extrapo-late or make assumptions and guesses, is a powerful analytical tool.

Coloured pencils

Among the most unusual neurological conditions I encountered among colleagues were prosopagnosia and synaesthesia. Most of us at times struggle to remember faces, but prosopagnosia is a condition beyond this, amounting to 'face blindness'. It is, to quote the Apple co-founder Steve Wozniak, who suffers from it, impossible to make 'new memories of faces'. In some cases this cognitive disorder seems to originate in the part of the brain that normally treats human face recognition in a special way, differently from the processes for recognition of other objects. Because this function is impaired, prosopagnosics may even struggle to recognise themselves. They may not realise that this is unusual at all until relatively late in life, despite the difficulty it may present for them in forming relationships from school onwards.

One former colleague told me he had to learn ways of attaching memory tags to faces he needed to remember, including in his own family – a particular facial feature or a voice that he could remember as attached to a particular individual. One of the reasons a familiar and supportive workplace was so important was that the opportunity for offending people by not recognising them at all was substantial. But he was exceptionally good at other mental processing; perhaps because he had to work so hard to find ways of categorising and remembering faces, he had developed techniques for categorising disparate areas of data and information. And surprisingly, in a world where images can be easily faked, seeing every facial image as new can naturally come as an advantage.

The most fascinating neurological condition I came across, this time not in GCHQ but in another government facility, was synaesthesia. This trait above all others highlights how little we understand the way we perceive, sense and process information about the world around us. For synaesthetes, something they experience in one sensory area automatically prompts a reaction in another. Most commonly, a number may be experienced as a colour – always the same colour – so when someone sees '3', they experience it as yellow. Likewise a sound – a violin, say – may automatically appear as the colour blue.

Those with synaesthesia may of course go through life assuming everyone experiences the world in this way, especially as it does not prevent them from reading and writing, or working with numbers. But the condition confers some clear advantages. The categorisation of numbers as colours may help in calculation – some synaesthetes certainly seem quicker at mental arithmetic than the rest of us. But above all, the ability to see information in one form represented in a completely different one goes to the heart of data visualisation.

Data visualisation is as key to intelligence work as it is to any form of advanced problem-solving or statistical modelling. We are much more likely to crack a complex problem if we can see it laid out properly, as a whole, in all its complexity. To take a very basic example, Welchman's use of coloured pencils at Bletchley Park to label the different radio networks was so crucial to him that he ran out of British supplies and had to get pencils shipped in from America. Interviewing for new recruits, he asked whether they were colour-blind, and later recalled:

It may seem absurd, but these colours really helped, and we could never think of an adequate substitute. Men from the machine room ... would keep wandering in to examine the traffic charts, and the identification of keys by colour was a great help to them.

In synaesthesia, the stimulation of a particular sense, sight, for example, leads to instant and involuntary reactions in another neural pathway, such as sound. The traditional theory is that different parts of the brain that should not be talking to each other somehow leak. Recent research, however, such as the work of neuroscientist David Eagleman, suggests an alternative theory to explain this cross-fertilisation. He argues that the 'plasticity' of a synaesthete's brain is unusual, lacking the ability to modify an association once it has been set. According to this hypothesis, a child with synaesthesia whose first sight of the letter 'J' is a purple letter 'J' on a primary school wall will have both letter and colour fixed in the brain, while other children will go on to see 'J's in lots of different colours and begin to modify the automatic association between the letter and a particular colour.

Different ways of making sense of the world may be extremely useful in an organisation, while having some consequences for the individual concerned and for those around them. At Bletchley Park, in Room 40, or across the Atlantic at Riverbank Laboratories or in the Black Chamber, the inherent value of individuals who thought differently meant that wise leaders – or those who understood enough to know what was needed – found ways of not simply tolerating but actively using these staff.

In an age when neuroscience as a separate discipline was still in its infancy, these different ways of perceiving, sensing and processing tended to be lumped together as 'eccentricity', as defined by the manifestations of the condition when seen from the outside. Nurturing this behaviour, and allowing the individuals involved to flourish and be gently directed towards the task at hand, was a key part of the genius of Bletchley Park and its successors.

Caves and marketplaces

Like most modern employers, GCHQ has formalised the tolerance and support it always showed for unusual behaviour in 'reasonable adjustments', to use the language of the statutory world. These techniques themselves have been tinkered with and adapted inside GCHQ, as one might expect of staff who instinctively like to look under the bonnet. Software to help 'map the mind' or set out thoughts visually, voice-to-text programs and more basic techniques – even Welchman's coloured pencils – have helped neurodiverse staff translate their own experience of the world for themselves and for others.

The kind of recruitment and promotion interviews for candidates or employees common in most organisations unintentionally heighten anxiety among many neurodiverse people because the questions are unstructured and free-flowing. They essentially test social interaction and presentational skills, and favour quick-firing, articulate bluffers. At GCHQ, interviewing, appraisal and all the paraphernalia of civil service bureaucracy have been re-thought through a neurodiverse prism.

But alongside practical support, the atmosphere and the buildings matter too. The shift from the old post-war headquarters in Cheltenham to the brand-new Doughnut in 2003 was a huge culture change, with an abrupt transition from the privacy of small rooms to a vast and public open-plan arrangement. Moving to open-plan was highly innovative for a secret agency in which information is 'compartmented' – very few people need to know everything that is happening. But it was also a particular shock for some neurodiverse staff who were used to their own small and solitary space.

Significant efforts were put into creating quiet spaces, an ever-increasing challenge as 9/11 and the rise of Islamist terrorism led to a rapid expansion of all the intelligence agencies. These were not, of course, new problems. The Huts at Bletchley Park were manically open-plan and often noisy, with limited separate office space and paper-thin walls (as well as minimal heating). Turing himself would come in late at night to pursue his own mathematical research at a time when he could find some peace. In fact, much thought has been given to the physical space needed by mathematicians in subsequent GCHQ buildings, as well as in academic institutions – a mixture of 'caves' for solitary reflection and social marketplaces for the exchange of ideas. Architecture matters in an organisation trying to enable cognitive difference.

Conventional estimates of how many people nationally are on the autistic spectrum range from 1 in 500 to 1 in 1,000. Informally we used to guess that 1 in 4 of GCHQ's workforce might have some neurodiverse condition, in many cases mild and not diagnosed as autism, but the proportion is still striking. It should be stressed, however, that the organisation did not intentionally recruit a certain percentage of neurodiverse staff, nor tie a particular condition to a specific task or function; it would be daft to suggest that 'X is a job for a dyspraxic' or 'Let's get someone with Asperger's to do Y', which would both over-estimate our understanding of neurodiversity and oversimplify its impact on the individual. Understanding and employing the unique talents of any given individual is what makes this difficult and interesting as a human resources task – indeed it challenges many of the current approaches to recruitment and reward.

The real lesson of the secret world is, first, that neurodiverse individuals supply something extra, which, when mixed with the contribution from non-neurodiverse employees, produces something otherwise unachievable. It is not possible to predict in advance exactly what this chemistry will produce, but the risk is worth taking. Neurodiverse talent is hard to measure and fit into a recruitment template, and needs more support than an organisation might normally want to give. This perhaps explains why it has not been sufficiently valued elsewhere in the past.

More importantly, the rest of us should aspire to some of the different ways of perceiving that neurodiverse staff have by birth and brain wiring; in short, we need to adopt some of the cognitive approaches of autism and challenge our own 'normal' ways of thinking.

The good news is that recent neurological research suggests that all our brains exist at some point along the spectrum of these 'disorders', because the brain is plastic, constantly changing and repairing itself. The traditional 'autistic spectrum' may in time be seen as one end of a much broader human spectrum. Many of the conditions outlined above will resonate because most of us will have experienced some part of them, albeit in a more diluted form. There are a number of self-tests for autistic traits approved by health authorities; take one, and you may be surprised to learn that you indeed experience, or have experienced, some of these characteristics.

So it follows that if the majority of organisations were to look a bit more closely at their staff, shorn of the organisational and cultural straitjacket, they would be surprised by the scale of neurodiversity concealed within the average workforce.

But the real discovery, accidental perhaps, in the secret world has been the benefit of a mixture of these approaches in a single team. Blended together, different approaches to a problem set yield results impossible to achieve with a single individual. This is the key difference at Bletchley and in the years since. Rather than seeing cognitive difference as a handicap to be put right or hidden away, it is something that is uniquely valuable when added to the mix.

19

First Principles:
Antidotes to Bias

Fans of *Homeland*, the TV series in which CIA officer Carrie Mathison thwarts various threats to the United States, from Islamist terrorism to right-wing extremism, will be familiar with a central theme in the show that breaks new ground for a spy thriller. Carrie suffers from bipolar disorder, and Claire Danes, the actor who plays her, brilliantly depicts what the symptoms of this illness are like. In much of each season we see Carrie in a medicated steady state, but when she is 'off her meds' we get glimpses of her highs and lows. Her manic or psychotic phases involve huge bursts of energy and hyperactivity, sometimes expressed in extreme happiness or in anger and frustration; she sleeps very little. The depression that follows these phases involves despair, self-harm and days in a darkened room. I am told by those who understand the illness that Danes's portrayal is extremely accurate, if occasionally heightened for dramatic effect.

Looking objectively at Carrie's condition, a steady career that enables her to manage her medication would be an obvious choice to keep her on an even keel. An affair with a US Marine who is a potential suicide bomber, followed by the personal pursuit of an Iranian intelligence chief, an Afghan warlord and a right-wing conspiracy within the US intelligence community look less than ideal, a point that Carrie's sister and other long-suffering friends make from time to time.

But the serious issue that the drama touches on – and one highly relevant to intelligence work – is Carrie's own belief that her hyperactive phases are cognitively significant; she has a heightened sense of awareness, and begins to see patterns and solve puzzles that her steady-state self – and other people – would not. We see manic wall displays on which she has frantically linked individuals and incidents to piece together a terrorist plot. She feels that her condition gives her an edge in seeing the world differently.

There are of course some important caveats here. First, it is easy to romanticise bipolar disorder and other conditions; most of those who experience the extremes of mental illness that Carrie displays will often be extremely distressed and will seldom solve world crises. Even Carrie wonders whether she has her priorities right or whether she should settle for a more 'normal' working life, although she concludes that she is good at intelligence and the sense of purpose keeps her going. More importantly, while the cult of the individual genius makes good drama, it is rarely true in the real world of intelligence. As we have seen, even Alan Turing was a team player and his achievements were assisted by others.

At the heart of *Homeland*'s storyline is the idea that neural difference can be an asset. Unmedicated, Carrie's brain processes information in new and different ways, and at high speed. This being drama, the insights she gains as a result always turn out to be correct and are almost always unique to her. While unlikely to be true in reality, the show's central thesis, that neural difference and cognitive diversity – crudely, different ways of perceiving, experiencing and thinking about things – have unique contributions to make, was and remains central to the success of Bletchley Park and GCHQ, as I trust I have shown.

But there is a more particular characteristic of the neurodiverse, rendered extreme in the character of Carrie, that is vitally important in many different types of organisation, namely the refusal to submit to groupthink. If the compulsion to systematise or seek patterns is one typical aspect of neurodivergent minds, then the tendency to find human interaction and emotion difficult is another. Coupled with compulsive and rigorous attention to detail, this makes a uniquely powerful contribution to the world of intelligence. It means that going with the flow, conforming to the consensus or toeing the line are felt as alien and difficult. This may be interpreted as stubbornness, and sometimes as obsession, but it is uniquely valuable.

Swimming against the current

The greatest intelligence mistakes of the past have been caused by the natural cognitive biases of analysts. Confirmation bias – hearing what we want to hear – underlay a major part of the conclusion that the Iraqi regime under Saddam Hussein possessed weapons of mass destruction, as did choice-supportive bias, the tendency to stick with one's original judgement despite growing evidence to the contrary. These personal prejudices, and many other forms of bias, are baked in to most of us and are hard to resist, whether at an individual or institutional level. Challenging them demands both an obsessive return to first principles – looking at facts as if you have never previously seen them – and a willingness to swim against the current.

While most of us make assumptions, fill in gaps and read between the lines, or readily move to broad conclusions, many neurodiverse people find this intrinsically difficult.

In a highly charged atmosphere, perhaps after a terrorist atrocity, the ability of people with neurodiverse conditions to be dispassionate and assess evidence purely on rational grounds, excluding emotion, can confer a huge advantage. One of the reasons why conspiracy theories flourish is the powerful emotional drivers of fear and paranoia. More commonly, emotions skew the assessment of statistics: we see what we need to see. So the ability to look at a complex and sometimes person-ally affecting investigation as a dispassionate intellectual problem – a puzzle – can be a great asset to a team. To do so may be perceived as 'cold', but it is invaluable, the antidote to the natural emotional biases of others and a powerful challenge to groupthink.

In Carrie Mathison's case, her obsessive convictions about conspiracies happen to be correct – they are rational and based on facts that are too difficult for others to see or connect; and the consequences of her being right are simply too profound to contemplate.

But obsessions can also be wrong. The McCarthyite purges of the 1950s in the United States were mirrored by a zealous campaign within the intelligence community led by James Jesus

Angleton, the legendary CIA counter-intelligence chief. The fact that he was duped by Kim Philby and others – and perhaps the level of personal betrayal he felt – led him to suspect KGB infiltration everywhere, to launch destructive purges and to doubt key evidence to the contrary from dissidents. He was at the conspiracy theory end of the saner approach taken by Hugh Alexander in the Venona project, as earlier discussed in Chapter 9.

Angleton's problem was that he had allowed an emotionally driven narrative or hypothesis of widespread systemic betrayal to colour all subsequent analysis. He could no longer look at anything from first principles, and simply slotted emerging facts into a grand, pre-ordained conspiracy. He did huge damage to his own organisation, on a scale that his enemies could only have dreamed of.

The failure to be dispassionate and analyse from first principles is ultimately disastrous in an intelligence context. When Sir John Chilcot's Iraq Inquiry finally reported in 2016, one of its key conclusions was that the intelligence agencies produced 'flawed information'. Because Saddam Hussein had possessed and used chemical weapons some years earlier, these agencies simply discounted the possibility that he might have abandoned them.

So powerful was the ingrained cognitive bias that evidence to the contrary was explained away. Most famously, a secret human source relayed to MI6 an account of Saddam's chemical nerve agents being carried in glass beads. To anyone who had seen *The Rock*, the 1996 thriller starring Nicolas Cage and Sean Connery, this ought to have rung alarm bells, especially as such lethal chemicals were not usually carried in this way, or at least not outside Hollywood's imagination. Of course, it did ring bells – MI6 watched movies too and were aware that this might be a source fantasising or telling his handlers what they wanted to hear – but they nevertheless gave the information significant weight.

Subsequent inquiries into the intelligence failures of the Iraq war reached many of the conclusions that will be familiar from the preceding chapters of this book: analysts failed to listen to experts, understood too little of the social and cultural context of Iraq, and failed to test sources. Swayed by deep-rooted and unexamined assumptions, they were carried along by the consensus and were too susceptible to conformity.

Finally, at the risk of adding a further layer of intrigue, understanding these human failures and how they also affect your enemies can be a powerful weapon in reverse. Published documents recovered during the raid on Osama bin Laden's compound in Pakistan suggest that paranoia in a terrorist group can be as powerful as infiltration. Organisations that are convinced that they have been systematically compromised begin to purge themselves without any help from outside. Ingrained assumptions of treachery are as dangerous as lazy complacency.

The only antidote is to be rigorous, and look at information rationally and free from bias. In that, stubborn and 'awkward' people – even extreme people – who find conformity difficult, may be a crucial asset in the mix of skills.

The engineering mindset

Working from first principles is helpful not only in analysing data; it is central to the engineering mindset, and is the problem-solver's, innovator's and inventor's key characteristic. Simon Baron-Cohen, who has spent a lifetime working with autistic people and studying autism, has noted the overlap between famous engineers and neurodiverse conditions.

In recent years the public narrative of Bletchley Park has shifted its focus away from eccentric dons to a more psychologically subtle portrait of the solitary neurodiverse genius, with Turing being confidently described as autistic, particularly after his portrayal by Benedict Cumberbatch in *The Imitation Game*. But neither of these depictions is quite right.

When Dermot Turing, Alan's nephew, visited GCHQ to launch a memoir of his uncle, he was asked about the popular attribution of autism or Asperger's to the great codebreaker. Dermot was clear both from his shared family memory and from his study of Alan's life that he was 'socially awkward' but not autistic.

It is true that as a boy, Alan Turing was more comfortable with numbers than words; his spelling in later life could be erratic and his writing worse. He apparently suffered from left–right confusion, an inability to distinguish the left and right hands; this is a common condition, and there has been considerable and related research into hand–brain coordination, but none of the findings point to anything unusual in Turing. The search for a neurological explanation of his extraordinary mind is unlikely to come up with anything now, but it would surely have interested a man who from childhood had been preoccupied by the brain as a machine. His later work, of course, increasingly focused on the 'wiring' of the brain and how that might be computable.

A number of Turing's colleagues speculated about the nature and origin of his intelligence, although none of them doubted his genius. Jack Good, a brilliant young colleague of Turing's, identified his thought process as his key characteristic:

He liked to start from first principles and he was hardly influenced by received opinion. This attitude gave depth and originality to his thinking, and also helped him to choose important problems.

Good eventually established a close relationship with Turing (who was at first unimpressed when he found the new recruit asleep under a table during his shift); Good taught him the ancient Chinese game of Go, a significantly more complex challenge than chess. He regularly beat Turing at Go and, like most of the expert chess players at Bletchley, considered Turing a rather average chess player. Unlike most advanced players, who work off established sequences of moves, Turing's tendency to look at each possible move from first principles set him at a distinct disadvantage. Significantly, Good said of Turing that he 'thought deeply rather than quickly and [Turing] said his IQ was only about the average for Cambridge undergraduates'.

The older and grouchier Dilly Knox also admired Turing but commented in some exasperation:

Turing is very difficult to anchor down. He is very clever but quite irresponsible and throws out suggestions of all sorts of merit. I have just enough authority and ability to keep his ideas in some sort of order and discipline. But only just.

This was rich, coming from Knox, but many of Turing's fellow workers also commented on his extreme difficulty in communicating or explaining ideas. This is why it is unlikely that Turing would excel at many of the entrance tests for today's public and private sector organisations, which tend to prize speed and the ability to process, precis and present with confidence large amounts of information. Turing and his colleagues, by contrast, encouraged an approach that valued reflective thinking about problems, without the prejudice of easy assumptions.

In January 1940, before the mechanical bombe machines were manufactured, a twenty-two-year-old Northern Irish mathematician arrived in Turing's section. Instead of focusing on the mathematics – on which the team was stuck – John Herivel sat back and thought about the very human behaviour of Enigma machine operators. He took as his hypothesis that some code clerks would be lazy, and that these lazy clerks would take particular shortcuts when setting the machines' rotors and rings each morning. If he was right, it would be likely that not all, but a reasonable proportion of clerks around the world would be similarly lazy, thereby making it possible to identify this cluster by plotting all the first messages of the day on a grid. He was correct. Without this one insight Bletchley would have lost visibility of enemy communications during the key months of 1940 when Britain was at its weakest, a point no doubt underlined when Herivel's senior colleagues arranged for him to be presented to Churchill.

But it was Turing's approach to the physical world, brilliantly explored by his biographer Andrew Hodges, that is the key to the difference between him and many others. What is often seen as Turing's eccentricity can be explained by his relentlessly scientific approach – a natural determination to start with first principles, and to link the physical and the intellectual. We have already seen his fascination with natural patterns and the Fibonacci sequence. Turing is a good example of someone whose strangest behaviour always had a rational explanation and was never affected, although he had a great sense of humour and enjoyed winding up the pompous.

Peter Hilton, who worked with Turing in Hut 8 at Bletchley and afterwards at Manchester University, gives a number of examples. Contrary to the professorial stereotype, Turing was a fine athlete, a long-distance runner who but for injury would have been a serious contender for the British team at the 1948 summer Olympics in London. At Bletchley he was a keen tennis player, but became dissatisfied with his volleying performance close to the net in doubles. He analysed this as follows:

The problem is that, when intercepting, one has very little time to plan one's stroke. The time available is a function of the tautness of the strings of my racquet. Therefore I must loosen the strings.

Hilton recalls that Turing slackened his own strings and turned up for the next match with 'a racquet resembling a fishing net. He was absolutely devastating, catching the ball in his racquet and delivering it wherever he chose – but plainly in two distinct operations and, therefore, illegally.'

It is significant that Turing made the changes himself – he was not solely confined to the world of the mind, and liked to roll up his sleeves and make things with his hands. From his childhood until his untimely death he constructed his own devices and carried out experiments at home, with varying degrees of success (he came close to burning his house down).

Other stories that Hilton tells about Turing reflect both the affection in which he was held and confirm his consistently scientific way of looking at the world. Like all his contemporaries he cycled everywhere, and at one time found that his chain was coming off regularly; by counting the number of revolutions of the chain ring between each occurrence he was able to pre-empt the malfunction without stopping to put the chain back on and get his hands oily. The sight of this must have been unusual, the more so in summer when he would go out wearing a gas mask, which he thought, correctly, would alleviate his severe hay fever.

Not all his practical projects were a success. Against what seemed a reasonable likelihood of German invasion, he invested £250 of his savings (around £15,000 today) in silver ingots, which he transported in a pram and buried in two locations around Bletchley. He was right about silver holding its value after the war but wrong about his ability to find the hiding places in a changed landscape. Despite searching for three years, he never found his ingots.

Practicality – or the practical construction and application of solutions – is the domain of engineers. Quite by chance I have spent much of my career with them, from military engineers to software developers, data centre cooling specialists, satellite communications experts and cyber security practitioners. I have had the great good fortune to get to know some spectacularly talented engineers, none more so than Vint Cerf. One of the creators of the internet, in the early 1970s he devised with his colleague Bob Kahn the protocols that govern the transmission of all our data. He went on to create the first email service across the internet, along with many other innovations.

A mathematician by background, he was attracted to the practicality of purposeful solutions:

Engineers solve problems and that's what I like to do. I like to make things that work that other people can use. There's nothing more satisfying than working really hard to make something work and then see that other people find it useful too.

Many of the core group of internet pioneers were close friends. Their story has often been told, but in the context of Bletchley there are interesting parallels, especially in their culture and the tasks they were set. As Cerf notes, some of the toughest problems were left to them as graduate students to solve: 'You just give engineers problems and say go solve them.' They rose to the challenge and, just as we witnessed at Bletchley, were 'too young to know you can't do that'.

Over the last twenty years GCHQ has rediscovered the role that engineers played at Bletchley, particularly Tommy Flowers of the General Post Office and Harold 'Doc' Keen of the British Tabulating Machine Company. Without these people and their teams, neither the bombes nor the computers would have been built, although they were not rewarded or recognised by the state during their lifetimes in the way they should have been. This reflected both an antiquated and slightly snobbish approach to engineering as a profession, combined with technological ignorance in government circles.

Flowers was bitter about the destruction of the Colossus machines and the secrecy subsequently imposed. He watched the ENIAC computer in the United States being unveiled shortly after the war and 'had to endure all the acclaim given to that enterprise without being able to disclose that I had anticipated it'.

The recognition of Flowers and those who built and ran the machines at Bletchley is an important part of the much larger project of changing attitudes to engineering. As we have already discussed, the failure to promote STEM subjects and their integration with other disciplines has put the UK at a significant disadvantage, a trend that continues. The Organisation for Economic Co-operation and Development's annual report shows that the UK has a shortage of construction, engineering and medical skills. Academic research continues to be a strength but the application of that study lags behind what is being achieved in other countries. To the extent that Bletchley and its predecessors were successful – and GCHQ continues to be – it was largely because they fostered this ability to do the practical application as well as the study. Whether it was Patrick Stewart in the 19th-century world of Faraday and Maxwell, or Turing in his own brilliant academic universe, these were people who tried things out, failed and tried again.

A final footnote on what links 'difficult' people, the uncompromising behaviour of Carrie Matheson, and engineers. A study published a few years ago looked at the academic backgrounds of Islamist terrorists over many decades and found that by far the largest proportion – at 44 per cent – had studied engineering. Medicine and religious studies were also represented, but the humanities hardly at all.

There are a number of possible explanations for this, not least the fact that engineering is an important resource for this group. But in the light of the cognitive approaches discussed in previous chapters, the psychological mindset required for excellence in engineering is worth considering. The authors of the study suggest a possible overlap between the desire for structure, order and certainty, or 'cognitive closure', that has been identified as a key personality trait of extremists, and the desire for structure and the optimisation of a clear process with no room for fuzziness that are often seen as the qualities of good engineers.

I am not convinced that there is any such correlation, nor am I, of course, really implying that engineers are prone to extremism – although I have often enjoyed suggesting this hypothesis to engineering colleagues – but at the very least their professional working methods may highlight the discomfort they feel at ill-defined, ambiguous solutions, or confident assertions that go beyond their expertise. Engineers are quite rightly professionally programmed to prefer clarity and challenge imprecision, and this is potentially a great analytical advantage.

Whether at the level of undisputed mathematical genius, working engineer or humble intelligence analyst, the ability to approach problems from first principles has always been a huge advantage in the secret world. It does not follow that those who do this are always right, but the challenge to received wisdom and the views of the majority machine is the key. And if that involves some difficult or even extreme behaviours, the job of the organisation is to manage and re-direct that.

20

Boiling Frogs, Disrupting Ourselves

'The achievement, it was true, hung by a hair.'

Nigel de Grey on Enigma decryption, 1940

Around the time that our intrepid Victorian cable layers were spreading their pre-internet network across the world's ocean beds, an MIT professor was boiling frogs alive.

For those who are not familiar with the metaphor, the idea is that if you put a frog in boiling water it will jump out. If, on the other hand, you put a frog in cold water and heat the water very gradually until it boils, the frog will acclimatise to the slow increase in temperature, will fail to jump out and will die.

This interested researchers into animal reflexes, and in 1888 William Sedgwick, professor in the Department of Biology at MIT, published a paper that attempted to settle the dispute about whether frogs would try to jump out as the water temperature rose. His conclusions failed to prevent further research, with the consensus among modern biologists being that the boiling frogs theory is not true.

Sedgwick went on to play a significant role in American public health, particularly in promoting clean water supplies, and he is sometimes regarded as the country's first epidemiologist. He would probably be a more controversial figure on the MIT campus these days given his extreme views on difference between the sexes; he argued on biological grounds that equality would mean 'a reversal of the whole social evolutionary process', setting back the world 'a thousand years'. His saving grace was, perhaps, his equally unflattering remarks about men.

The boiling frogs metaphor has been used – some would argue, overused – to describe any gradual change that is recognised too late, from climate change to conspiracy theories about the erosion of personal liberty. It is particularly apposite to describe the dynamic challenges of complexity and disruption posed by operating in the world of data and communications technology – and the challenge of transforming an organisation in the digital era. The pace and complexity of change make traditional 'big delivery' models less effective than ever: too slow, too sluggish and, simply, too big.

GCHQ used the metaphor in a paper published a few years ago entitled 'GCHQ: Boiling Frogs?', which considered how to implement complex organisational change in a disruptive technological world:

The pace of disruptive change is increasing, from the rise of cloud technology, social business, the Internet of Things and others. We need to jump out of our world and consider the big picture.

The paper attempted to summarise some of the secret world's approaches that have worked. The fact that it was published on an open-source tech site, GitHub, was itself significant. A secret technology organisation was explicitly recognising that a whole range of disciplines – from organisational change to software development – are best done in collaboration and shared with others. This was entirely consistent with the academic and engineering culture of Bletchley, now shorn of some of the secrecy imposed during the Cold War.

A century of disruption

The secret world that emerged in the 1950s made use of the power of computing it had helped to create, becoming one of the few parts of the British government that could preside over software development and engineering on a large scale. It also benefited from the space race through the advent of satellite communications, and was a key asset in the West's struggle against the Warsaw Pact. But the Cold War was fundamentally a static period for GCHQ, partly because the Soviet threat was monolithic and all-consuming, with no challenges of a similar severity developing elsewhere, and partly because communications – the signals – still followed pre-internet, single and identifiable routes around the world. Intercepting these was analogous to the cycles familiar at Bletchley Park – from Y stations picking up messages all the way to the finished intelligence product that emerged from the Huts.

As the Cold War drew to an end, the disruption to GCHQ's world came from two directions. First, the nature of what its customers wanted changed. Operating with fewer resources as government clawed back funding, GCHQ was now being asked to look at diverse threats in an increasingly volatile world: from the proliferation of nuclear weapons in developing countries to the rise of Islamist terrorism, serious crime, and ground wars in Afghanistan and Iraq.

But the far greater disruption came from technology itself. The internet fundamentally changed the environment in which all signals intelligence agencies – and the entire secret world – had to operate. The move to internet protocol technology meant a world where every communication – from emails and phone calls to web browsing – was broken down into 'packets'. Although these were fired around the world on the same under-sea cables, now fibre optic, the packets took different routes until reassembled at the destination computer. The global network and routeing changed constantly, mostly dictated by commercial factors, adding a whole new level of complexity to the challenge of interception.

Just as significantly, innovation in technology was moving out of the hands of governments. The internet both enabled and was enabled by the explosion of affordable computing, as semiconductors – 'chips' – became ever cheaper and more capable. The arms race of interception and encryption we had taken part in since the 19th century now suddenly changed; governments no longer had to compete just with other governments but with a global IT industry investing trillions of dollars in research and rapid development. Suddenly, they were better than nation states at large-scale innovation in this new communications technology. GCHQ, which had inherited a history of inventing and building its own technology – sometimes accompanied by an attitude that 'if we didn't build it, it isn't worth having' – faced being left behind in this race of asymmetrical scale and funding.

But change in the technical landscape also brought a more radical challenge for a secret agency. Within a short space of time most of GCHQ's targets were using the same technology and operating on the same platforms as the rest of the world. In practice this meant that those GCHQ was trying to tackle were going to be using the most secure communication available – based, ironically, on the public key encryption discovered in Cheltenham – that had been developed by the private sector. Worse, they were going to be 'hiding in the noise' of legitimate users.

Within two decades of its creation, the internet would present the possibility of mass-scale attacks – cyber attacks – against the entire economy and the whole of society, way beyond the far more local targets of old. The job of national-level cyber security inevitably gravitated to GCHQ for the reasons that codebreaking had gravitated to Bletchley Park: here were people with the right skills and the imagination to innovate solutions.

The task for GCHQ from the late 1990s was to find ways of developing technology that would keep up with the pace of change. A key decision was to adopt another Bletchley principle – using whatever piece of kit from outside that works – and adapt technology and business practice from the private sector.

This meant buying off-the-shelf solutions and deciding at which particular points it was necessary for GCHQ to invest its own expertise and resources for bespoke innovation.

007, licensed to programme (and project) manage

In GCHQ's case, a realisation of the scale of the challenge, combined with some financial mistakes in the 1990s and the headache of delivering a new headquarters building, led the organisation to place huge emphasis on programme management. This is probably the least glamorous task within any organisation and a million miles from James Bond, yet it is what has earned GCHQ the reputation of being the only part of government that can consistently deliver large, complex IT projects on time and on budget.

Programme management is arguably the ultimate challenge in practical, complex, dynamic puzzle-solving. As we have seen, Gordon Welchman grasped this in driving the Bletchley production line. Anyone who has tried to manage the smallest home improvement project of their own instinctively knows the problems of inter-dependency. What can be done in what order, the availability of the right skills and materials, and a range of other unpredictable factors, including costs, illustrate on a micro-scale the challenge of programme and project management. Scale this up to a series of hugely expensive, interlocking IT projects, involving multiple suppliers and fundamental change to the organisation – especially where demands for what the project must achieve constantly change – and it is easy to see why large-scale government IT ventures often fail.

The instinct of most organisations when confronted with an extremely complicated problem is to draw up and follow a set of instructions based on past experience. This may work up to a point, but it will limit the solutions to those we have seen before. Where a problem is highly complex and novel – as it was at Bletchley Park or indeed during a pandemic – there will be few such instructions. The problem and its interdependencies are new: understanding consequences, cause and effect, is difficult.

As we have seen, the political or business leader's natural desire to command and control a huge single project from the centre – to issue instructions and pull levers – is unlikely to work here. This has implications for the way we build an organisation, especially where it is technology-based, and the way we change existing organisations.

Change in any large organisation is difficult, the more so if it has enjoyed consistent success, which brings its own inertia. In such cases, most organisations will have started to arrange staff according to existing processes rather than their own particular skills and expertise, with the structure increasingly reflecting the way that work has been done in the past rather than the demands of the work itself, let alone what the future might hold. The more established an organisation, the more likely it is to look inward and worry about its own metrics.

In short, a standard hierarchical organisation will tend to fall victim to 'Conway's law', namely that organisations are constrained to build systems that mirror their own structures. It is always hard to step out and see from the outside what needs doing or what will work, but an organisation that contains a number of cultures, groups and organisations in tension may fare better, not least because it is much trickier to simply mirror itself. The desire for hierarchical symmetry and tidiness is a disadvantage in the pursuit of agility and speed in response to complex problems.

The disruption of cyber: becoming public

This capacity to change has been tested more than ever by the demands of cyber security. We have already seen that off-the-shelf technology, ubiquitous encryption and the mere fact that everyone – good and bad – is using the same technology platforms will present particular problems to the secret world. The examples of election interference by the Russians and the success of ISIS propaganda illustrate the asymmetrical use of this new world by adversaries. But it is the rise of cyber attacks over the last twenty years, and what to do about them, that more than anything have upended the assumptions of agencies like GCHQ, throwing the traditional role of governments to protect its citizens into doubt.

This in turn has caused intelligence agencies to rethink the benefits and limits of secrecy. There could hardly be a more extreme example of self-disruption and internal cultural challenge, a bit like the Royal Navy asking itself whether ships are still useful. The drivers forcing this step into the public space are many: concerns about the invasion of privacy mean that secret government agencies must be much more transparent if they are to secure public consent, especially as artificial intelligence reveals new worlds of possibility; technology itself is progressing in the open – software developers need to share techniques; and the epicentre of development is no longer governmental.

As we have seen, this trajectory towards openness is not new. From the blanket of secrecy spread over Bletchley that stayed in place for more than thirty years, GCHQ itself only emerged fully in 1994, when new legislation officially acknowledged its existence, although Mrs Thatcher was forced to partially lift the veil in parliament after the conviction of Geoffrey Prime, who spied for the Soviets at Cheltenham. Even then GCHQ remained the most secret and unknown of the three agencies, despite being larger than both MI5 and MI6 combined.

But, as mentioned, the greatest disruptive driver for openness has been cyber threats – particularly the 'democratisation' of large-scale crime and fraud – simply because these affect the entire country and its economy in potentially devastating ways. Estimated global losses of $5 trillion to cybercrime over the next three years dwarf even some of the pandemic spending figures. It was the scale of this that led GCHQ – and other agencies around the world – to question their role and redefine the parameters of their secrecy.

By volume, the overwhelming majority of malicious activity in cyberspace – perhaps 80 per cent – is criminal. It is about making money and is therefore surprisingly predictable. Cybercriminal groups look for poor defences and easy vulnerabilities that can be exploited at scale. They are agile, creative and innovative, pulling in different technical skills and capabilities in a manner that mirrors some of Bletchley's organisation, albeit transposed to the global online criminal world. The remaining 20 per cent of cyber threats are largely more sophisticated nation-state activity deployed for political effect or to gain economic advantage: stealing intellectual property, for example, is a good way to shorten the research and development cycle. There is a growing overlap between criminal and governmental threats, because some nations will license or tax cybercriminal groups to operate, and sometimes use them as deniable 'proxies' for their own activity. Finally, a few per cent of the problem may be either 'hactivists' – those pursuing political objectives online – or bored teenagers engaging in a bit of digital vandalism.

All of these groups and their objectives existed long before the internet, but technology has simply enabled them to do malicious things at greater scale, more quickly, more cheaply and more anonymously than ever before. The volume of processing first tackled at Bletchley through Colossus continues to expand enormously.

In the 1990s – the early days of cyber attacks when there was not much to target – gathering intelligence and responding to cybercrime looked like a normal extension of GCHQ's secret role. I remember having a long and half-serious conversation in the GCHQ museum with a deep expert in cyber about the first piece of Russian malware reverse-engineered in Cheltenham many years earlier: how, we both wondered, were we going to preserve this piece of code as an exhibit for the future, and what would a museum of cyber security even look like? But as the scale of cyber attacks increased, the pressure on governments to do something grew in tandem.

The fundamental problem was a lack of skills and under-standing as acute as anything in early 20th-century government circles regarding the comparable cutting-edge technology of the day. While there was in-depth expertise in Cheltenham, anything else tended to be in the private sector. We needed – and still need – a revolution in skills as profound as that which occurred in the 19th century. As we have seen, one of the main drivers for engaging women and other under-represented groups in this technology was to expand the recruitment of technically profi-cient cyber security specialists.

In the meantime, government increasingly looked to GCHQ for answers on cyber security. Strictly speaking, its remit was limited to official networks, a job it had been given seventy-five years before. Alan Turing had in fact spent more of his GCHQ career in communications security than on breaking Engima, working on encrypted or 'scrambled' voice calls at Hanslope Park, near Bletchley, and at the Bell Laboratories in New York.

The challenge for GCHQ – as for many companies on the 'digitisation' journey – was to grow the knowledge and experi-ence base of its small pool of cyber experts. This meant getting the relevant staff out of the secret world and having them work in and with the private sector and among the wider population. The isolation of GCHQ as a secret island was no longer tenable or desirable.

The solution lay in the Bletchley model, though few may have realised it at the time. It was simply impossible to lead the nation's operational response to cyber security from behind the walls of a top-secret institution. Within the Doughnut lay the government's only grouping of world-class cyber talent, yet what was the point of being dominated by the constraints of secrecy when the cyber threat was mostly anything but secret?

Denniston and his colleagues would, I think, have been comfortable with GCHQ's eclectic, hybrid and self-disrupting answer. The creation of the National Cyber Security Centre (NCSC) in 2016 involved taking a relatively small group of GCHQ's deeply technical cyber experts and installing them in a brand-new office block in London – an incongruously smart venue for anyone more used to dreary government buildings. In true Bletchley style, the NCSC added other people with a diverse range of skills and backgrounds to this small core of technical talent: industry cyber specialists who understood their sector better than any intelligence officer; communicators well versed in getting their message across to the public (an ability rarely found in abundance within the secret world); and policy officials who could manage and lead the enterprise while navigating their way through the minefields of Whitehall.

The NCSC exists as part of GCHQ, which provides it not only with its core group of deep specialists, but also access to secret capabilities and data, as well as the awesome power of allies such as the US National Security Agency. But predominantly its work is open. In its own way it was a bold step for government; the easy option would have been to create a new civil service quango, easier still for GCHQ to have retreated into its secret world and hope someone else would do this. It was also expensive. At a final meeting with George Osborne, the chancellor at the time, on whether to set up the NCSC, I handed him a copy of the 1941 letter from Bletchley codebreakers to Churchill, pleading for resources, to which he famously responded with urgency: 'Action this Day'. While I was not wholly serious in my comparison of our respective circumstances, I certainly believe that the scale of the global cyber threat dwarfs the scale of the problems given to Bletchley, and some of the solutions will require equal creativity.

The difference is where the balance of responsibility lies. In the 1940s it was taken for granted that a government's job was to protect the lives of its people; while that is still the aspiration, protecting everyone from cyber attacks or the abuse of online platforms looks like an impossible task.

The challenge was, and still is, to work out in the cyber age what interventions would make a difference at scale. The response of a particularly brilliant GCHQ specialist to the question of 'What could we do at a national level to change this?' was to come back, very quickly, with twelve technical ideas for protecting the country at scale. This shift towards an 'active cyber defence' interventionism by governments and companies – taking the burden off individuals and not relying on their own technical competence for defence – looks like the best hope for changing the weather in cyberspace.

A glance back to the pre-internet cables of the 19th century helps us to remember that the technology on which these threats are proliferating is still in its early days and that it almost always takes time for safety to catch up with technology development. The reckless driving skills of Dilly Knox and others of his generation led to escalating road traffic accidents before the Second World War, even though far fewer cars were on the road. Speed and engineering prowess were the drivers of progress, with no one thinking to regulate the safety standards of cars or constrain those who drove them. But the proliferation of accidents caused governments and the insurance industry gradually to insist on mandatory safety standards, from seatbelts and child car seats to anti-roll bars and ABS, sometimes removing the freedom to choose from individuals. The same will happen in cyber security – safety will be built in by design. But in the meantime the ride will be rocky.

GCHQ's journey out of secrecy has enabled the NCSC to do things that would have been unthinkable for an intelligence agency even a few years ago. Calling out Russian, Chinese, Iranian or North Korean cyber actions and describing some of the details of their work goes against most of the instincts of the secret world, which tries its utmost to keep such knowledge to itself. If knowledge is power, letting the other side know what you know is usually seen as a mistake. But in a world where threats operate over a technology used by everyone – and where everyone is at risk – the rules surrounding information have to be different.

It may well be that in twenty years' time the majority of the work of Bletchley's successors will be in the open public space; ultimately their job is to go where the problem is. As attacks on democracy increasingly operate on social media platforms, accelerated and amplified by artificial intelligence, governments will need to work with others to understand and counter them when the companies cannot. To be effective, and to enjoy the consent of the people, much of this will have to be done in public.

Of course, there will always be capabilities that need to be developed and used in secret. In the popular imagination, offensive cyber attacks are more exciting than defensive cyber security. Like conspiracy theories, they are particularly alluring for journalists looking to make sense of the complexity of the cyber world. The idea of a big red button – or maybe a big red mouse – which can be used to 'switch off' or destroy something in a hostile country is attractive, even to some politicians. But in reality, cyber defence will always be more important than offence. An open economy and a democratic society will always be asymmetrically vulnerable to attack.

The United Kingdom was one of the first countries to publicly announce its offensive cyber programme, now the National Cyber Force, and it has also openly acknowledged its use of cyber weapons against ISIS and cybercriminal groups. Very little can be said about this, naturally, but it is worth noting some obvious differences between conventional and cyber weapons that undercut popular perceptions. First, a cyber weapon is not static – it is not like a missile that sits on a shelf. It requires deep knowledge of the adversary's networks, and these constantly change. Once a cyber weapon is used, understanding the consequences or what collateral damage there might be is difficult. The explosive power of a missile can be tested, but a cyber attack may spread far beyond the networks explicitly targeted.

Nevertheless, these weapons are proliferating. In recent years there have been attacks against power grids and oil production, water-treatment plants, hospitals and transportation, alongside the breaches of private companies that hit the headlines almost daily. It is easy to imagine the public panic that would be caused by a successful attack on water or power supplies, or food distribution.

These incidents give us a glimpse into a possible future. State cyber attacks are not new, but the risk of miscalculation grows as hostile nations feel less constrained. It is only a matter of time before a state launches an attack that, perhaps inadvertently, leads to real physical harm to human beings.

During the pandemic, the world's first instance of a death attributed to a criminal cyber attack took place. An ambulance taking a critically ill woman to a hospital in Dusseldorf had to be diverted to another facility much further away because the hospital had been disabled by a ransomware attack; she died before the ambulance could reach the alternative hospital. It is easy to imagine these kinds of attacks, which can take place on a large scale and may originate from either criminal or state actors, accidentally causing harm and death.

It is possible to recover from most attacks, but they are hugely disruptive and expensive, and they undermine confidence in all the things we rely upon for daily life. Our dependence on the digital world, initiated by Turing, Newman and his colleagues in the bizarre circumstances of the Second World War, has become a new front of global vulnerability. Predictions of a 'cybergeddon' or a 'cyber 9/11' may be overdramatic; a cyber winter is not coming. But until nation states find better ways of settling their differences, and unless we can raise cyber security standards to combat criminals, we will have to get used to increasingly severe cyber storms.

21

Travelling Lightly

'Everybody has won, and all must have prizes.'

Secret history of Hut 6 at Bletchley Park, quoting *Alice's Adventures in Wonderland*

When Elon Musk took over Twitter in November 2022 he reportedly told staff to expect an 'extremely hardcore' ethic and that they would have to work 'long hours at high intensity. Only exceptional performance will constitute a passing grade.' This is not untypical of tech bosses and indeed other industry leaders. Bill Gates described his younger self as a 'zealot' who disregarded weekends and holidays, and was critical of his less committed staff's work patterns, not least because he gave forensic attention to the comings and goings in the Microsoft car park.

Leaving aside the lack of empathy displayed by ultra-driven bosses (admittedly, as we have seen, lack of empathy can actually be useful in some contexts), the experience in the GCHQ world offers a more interesting and counter-cultural route to innovation and creativity. This is not to say that staff at Bletchley

Park, or indeed GCHQ, did not work hard and that drivenness is unknown. Facing a very real existential threat from the Axis powers, codebreakers worked extremely long shifts in difficult conditions, particularly during critical moments in the war, from the Battle of the Atlantic to D-Day. By the same token, a terrorist crisis, a hostage rescue, a major cyber incident or support to a military operation might entail long and intense hours at GCHQ, lasting days or weeks.

Some unusual values of the secret world

But three human characteristics feature in the world of GCHQ and signals intelligence that would sit uneasily in a book about the corporate work ethic of big tech: the importance of detachment from work, the value of humility and the critical contribution of strong outside interests. These are both a consequence of secrecy and enabled by a secret environment.

Detachment arises from the very obvious fact that you cannot take your work home if it is secret. The global pandemic challenged that assumption, obliging intelligence agencies to rethink what could be done safely outside a highly protected building. But nonetheless it remains true that the most sensitive information and operations are never going to be suitable for work from home, meaning that intelligence officers have to leave their work in the office. They cannot take it home physically, nor can they unload it on family and friends in post-work conversation.

Of course, this does not stop a mathematician continuing to mull over a problem at night, as the Clifford Cocks story about multiplying primes shows; nor does it mean that analysts who have seen traumatic events, or disturbing images online, can suddenly wipe them from their mind. But they have to take a break from what Musk would regard as the ideal work ethic.

GCHQ directors have the pleasure of giving long-service medals from time to time, which were always a wonderful opportunity to delve into the past. I remember one colleague who said he had begun work as a teenage apprentice the year before I was born. Now, more than fifty years later, he was finally retiring. Starting out as an engineer in a particular discipline, he continued working in the same field till the end, becoming a walking history of developments in that technology since the 1960s. I politely asked whether he would miss the place. His reply was quintessential GCHQ: 'Overall I've enjoyed my time here, but I have lots I want to do.'

There is a link here between detachment and humility that is peculiar to the world of technical intelligence. In a context where achievements cannot be talked about publicly, no one is motivated by public recognition. Bletchley Park is the ultimate example of this self-effacing approach, forty years passing before anyone was able even to speak of what they did. Many years ago I met a married couple who had not told each other about their secret wartime activities until the 1980s, even though they had both been at the Park at the same time. For some, this need for silence could cause genuine pain. There are many accounts from veterans who would have loved to have told their parents about what they had contributed to, particularly when they had been accused of not pulling their weight in the war.

Efforts have always been made to celebrate this collective success internally – even at Bletchley the Huts would mark breakthroughs with quiet events. A few of those involved were given public honours, although randomly and sometimes in inverse proportion to their contribution.

In recent years King Charles III, as Prince of Wales, established an annual secret awards ceremony at St James's Palace for the intelligence agencies. A formal but suitably unglitzy affair, the citations are crafted to give nothing away and there are no gushing acceptance speeches. But the recognition by Prince Charles as patron of the agencies, and by colleagues, in front of families, was very important and hugely appreciated – and a small sign that the secret world can also evolve. But ultimately the work has to be its own reward. And this being the public sector, financial compensation is never a serious incentive.

This humility, whether natural or enforced by secrecy, brings a number of benefits. First it discourages self-promotion and pointless competition; pushiness, prized in many outside recruitment processes, is at best wasted effort in this world. It is therefore an environment that attracts those who want to focus on the work itself.

More importantly, the inability to take personal credit facilitates a degree of collaboration that is difficult even in academic life. As Cocks put it in an interview:

> GCHQ is much more collaborative [than a university]. You don't stand up in a seminar and say, 'I've got this half-baked idea that I can't make work.' But at GCHQ you want to get

solutions and you don't really care whether they're yours or a colleague's.

Finally, job satisfaction has to come from both the work and the 'mission'. It has to be driven by stopping bad things from happening to others, and by interesting, diverting work – by the game itself. It is not always possible to set out in detail how an individual's contribution fits into the bigger picture, but even at Bletchley Park the best managers went to some lengths to ensure that people understood the scale of the wartime successes they were enabling through their often tedious and relentless grind.

Such managers were not easy to find. If the secret world of GCHQ has something to teach about leadership, it is that the qualities described above are naturally accompanied by a disinclination to lead and a distaste for megalomania. In a structure like Bletchley or GCHQ, the incentives to be a leader are usually outweighed by the downsides, mainly doing less of what you really enjoy. By definition, the best leaders in this kind of world tend to be reluctant – dragged into the internal limelight for as long as necessary before escaping to do something more important. Because administrative power is of little interest and a distraction from the great game, leadership is a chore. But as the reluctant Alastair Denniston recognised, someone has got to do it.

The ability to travel lightly in their secret work is accompanied by an approach to life that is embraced, as we have seen, by most of the great codebreakers. Most of them had lively and deep outside interests. Modern organisations strive desperately to encourage their employees to achieve a 'work–life balance', but their senior leaders tend to model the exact opposite: a narrow obsession with the work for its own sake. For GCHQ, such a narrow focus would actually undermine the multi-disciplinary, multi-coloured approach to the world from which the magic of creativity and innovation springs.

One of the most striking aspects of GCHQ and of Bletchley Park is that staff have traditionally pursued a dazzling array of interests outside work. At one level this was reflected on the internal social media platform. Every conceivable hobby was available in a sort of perpetual Freshers' Fair. There was an obvious bias towards engineering – cars, bikes of all sorts, and Lego – but also a strong showing of very high-quality musicians. 'Top Secret Brass' would perform at Remembrance Day and other occasions, and over the years excellent orchestras and choirs flourished.

But at another level, a quick look at the lives and publications of GCHQ's most eminent employees down the years reveals hobbies pursued to a remarkable standard. Hugh Foss was the first to examine and report on an Enigma machine in 1926 – a machine bought for £28, on the hunch that it might become important, which is still in the GCHQ museum, not least as an emblem of value-for-money purchasing. In between breaking Japanese codes at Bletchley he ran the Scottish dancing club and went on to write several books of dances after the war, some of which are still in use. Joan Clarke, sometime fiancée of Turing and codebreaker in her own right, who continued working for GCHQ until 1973, also had a parallel life. In numismatic circles she is regarded as the great authority on the sequence of gold coinage of James III and James IV of Scotland, on which she published many articles.

Search for Mavis Batey on Amazon and you will find more books on garden history and conservation than on Bletchley. It was for the promotion of gardening that she received an MBE, not for her extraordinary work in breaking German military intelligence and Italian naval communications. The University of the Third Age owes the creation of its British wing to a GCHQ cryptographer who was undeniably busy during the Cold War, but found time to pursue many interests. Others wrote guide books, local histories and caving manuals.

There have also been national-level hockey players and Olympic-level long-distance runners, including Alan Turing, and a good deal of dramatic talent, notably Frank Birch, who appeared as Widow Twankey in the West End and in nearly fifty films and TV programmes. And, of course, there were and still are puzzle-setters, crossword creators and chess players, many of them appearing under pseudonyms.

It would be easy to fill a book with the extra-curricular activities of Bletchley and GCHQ folk, and their collective contribution to the local community in Cheltenham over the decades has been significant. Of course, intelligence agencies have the luxury of public funding, but it would be difficult to dispute the idea that the staff's intellectual diversity and extra-curricular interests have fed into the success of their work. This creativity is hard to quantify, but it goes way beyond the common-sense notion that time away from work is vaguely therapeutic. It is part of the conundrum of a life that is at one and the same time objectively important and a grand intellectual game.

Recruits to GCHQ know little or nothing about what they are going to be doing and, after a long career, can say nothing about the intervening years. Any public recognition is minimal and will focus on anything except their secret life. Margaret Rock, one of Dilly Knox's all-female team at Bletchley, is a fine example. After her extraordinary achievements in that team – acknowledged as one of the key contributors in the struggle to break Enigma – she moved to GCHQ and spent the rest of her working life there, involved in the codebreaking challenge against the Soviet Union, work that is still highly classified. All we know of her publicly is thanks to family reminiscences: her letters to her brother and other family papers collected by Kerry Howard.

We know that she lived quietly with her lifelong schoolfriend Norah Sheward until her death in 1983. She spent her spare time drying rushes from the River Avon to pursue her passion for basket weaving. We know of her gardening and contribution to village life, and her affection for her extended family. But of the largest portion of her life, in hours at least – at work as a mathematician – we know nothing at all publicly. She encapsulates the humility of a secret life in technical intelligence and the distortion for outsiders looking in; what we can know of her is just a fragment of the whole picture – the rest is opaque. There is a greatness in this silence and humility that is totally at odds with the current age of hyper-communication, where every detail of life has to be documented and published. By contrast, these individuals who spend their lives quietly breaking the encrypted communications of those trying to do harm, remain silent.

Reflecting on where credit should lie for the great achievements in a key part of Bletchley's codebreaking, Hugh Alexander did not pick a single individual, even Turing. He reached a most un-modern conclusion. Quoting *Alice's Adventures in Wonderland*, he began with 'Everybody has won, and all must have prizes.' This quote was carefully chosen in a milieu in which Lewis Carroll, as puzzle-setter and mathematician, was far better known and admired than he is now. Carroll put these words in the mouth of the Dodo at the end of the Caucus Race, in which all the participants run around in any random direction until the race is declared over. It was intended, among other things, to be a satire on pointless political activity and posturing, but it was not a bad metaphor for Bletchley and GCHQ's approach: explicit recognition of the highly complex and sometimes chaotic team pursuit in which they were all engaged, but which was ultimately successful. A team game where personal credit was unimportant.

Conclusion

In Stanley Kubrick's 1968 classic *2001: A Space Odyssey*, HAL 9000, the on-board computer, famously refuses to allow the crew to switch 'him' off with one of the most mundanely chilling lines in film: 'I'm sorry Dave, I'm afraid I can't do that.' HAL has become so dedicated to his mission goal that everything else, even the lives of the human crew, are secondary and must be sacrificed if necessary.

Arthur C. Clarke, the co-author with Kubrick of the film's screenplay and the novel that appeared in parallel with the film, chose the year 2001 in homage to Alan Turing's predictions about artificial intelligence, published in 1950. In fact, Turing first wrote about 'machine intelligence' in around 1941 at Bletchley Park, although that work is lost. In breaks from his work on Enigma, he had been discussing with Donald Michie and other colleagues how machines might learn from experience, and exploring what a machine that could play chess might involve. It was not until 1997 that IBM's Deep Blue beat the world chess champion Garry Kasparov, and another twenty years before DeepMind's AlphaGo defeated the reigning world champion Ke Jie in the highly complex ancient Chinese board game Go.

Stanley Kubrick's adviser on *2001: A Space Odyssey* was Jack Good, the young mathematician whom Turing had found asleep under his desk during a shift at Bletchley Park. Sleep was important to Good; he stuck rigidly to shifts, not giving in to the temptation to keep on working. As a result, he once broke a code while dreaming, bringing the solution in to work the next morning. He went on to work for GCHQ after the war, then returned to academic life. As far as anyone knows, he is the only member of staff reportedly to have been honoured by Hollywood's Academy of Motion Picture Arts and Sciences.

Building on Turing's pioneering work on computers that might imitate humans – thinking machines – Good developed some key concepts of artificial intelligence. In an article in 1965 he described the concept of an 'intelligence explosion' or moment of 'singularity'. An ultra-intelligent machine, Good argued,

> *could design even better machines ... and the intelligence of man would be left behind ... thus the first ultra-intelligent machine is the last invention that man need ever make, provided that the machine is docile enough to tell us how to keep it under control.*

Appropriately for an adviser to Kubrick, Good questioned why his conclusion was seen as the stuff of fiction: 'It is sometimes worthwhile to take science fiction seriously.'

For all the hype surrounding artificial intelligence today, it is remarkable that the central questions about what it means for machines to think and what human consciousness consists of have not significantly changed since Turing and his colleagues explored them in the post-war years. Critically, these mathematicians and scientists had been deeply engaged in engineering and the physical application of technology, as well as being excited by the theoretical possibilities. They were also acutely aware of the ethical consequences of what might be created in their own area of 'intelligent' machines. In common with J. Robert Oppenheimer and members of the Manhattan Project, they could glimpse what the misapplication of technology might bring. But above all, the most gifted technologists had reflected deeply on what it was to be human; Turing, in particular, scatters insights on human life – a life that for him had been especially difficult – throughout his key paper on intelligent machines.

What they could not do at that stage was to predict the specific applications of AI, many of which are only now beginning to emerge. Alongside its huge benefits to humanity, such as in healthcare, education and transport, AI also brings threats, quite separate from the existential *2001*-style worries about computers taking over the world.

Artificial intelligence is already helping cyber attackers to scale up more quickly and enabling fraudsters to prey on their victims more convincingly. AI's ability to imitate and replicate human behaviour plausibly, at speed and on a vast scale, is also leading to a growing wave of disinformation, a further challenge to democracies.

At the heart of this is a problem that Turing and Good's successors in the academic and secret worlds will have to tackle: how to understand the huge troves of data on which AI models are trained. The greatest threat is that this data will be deliberately or accidentally manipulated or 'poisoned'. If an autonomous vehicle is made to believe that there is no brick wall where there is one, the consequences are obvious. But much more insidious will be data that leads to whole societies being duped or misled.

More parochially, AI and data are changing the very nature of the secret world. James Bond's business model is in deep trouble. As recent films in that franchise have shown, 007 is struggling to adapt to a world that is awash with data. His problem is not so much that he is a technophobe, but that the forces of the data economy are ranged against him. In the real human intelligence world, as well as on film, the old models are increasingly unsustainable. Turning up with a business card that states you are 'Richard Sterling' and working for 'Universal Exports' would no longer open doors for Daniel Craig in real life.

A brief internet search, never mind more sophisticated credit and social media checks, would quickly raise the alarm. If this weren't enough to unmask him, bulk data analysis certainly would. So simple Google searches or ChatGPT, let alone future intelligent machines, may be a greater threat to Bond than Blofeld ever was.

This is because every individual now grows up with a complex official and personal digital history, leaving a trace of their 'digital exhaust' wherever they go, most of it owned or captured by private companies. Going back and creating a fake or alternative digital life from birth would involve changing the records of an incalculable number of organisations and public databases. Even the most dyed-in-the-wool conspiracy theorist would find the possibility of doing this hard to swallow.

How then might today's heirs to Bletchley, whether in the secret or the open worlds, set out to tackle these colossal technical and ethical issues? Anyone looking for simple conclusions from history, or a prescription for innovation, creativity and problem-solving, is searching for the wrong thing. There was no masterplan, and perhaps that is the point. But there was clarity about the problem. There was an imaginative selection of personnel available to tackle it, operating within a structure to enable them to flourish. In the process, Bletchley identified people who were also very good at organising the chaos around them into an efficient production line. And from that line there was certainly output; these individuals were ahead of the curve in their creation of digital programmable computing, as well as in rethinking encryption and imagining artificial intelligence.

In looking for pointers from the past, the best we can do is to highlight some factors that run counter to received wisdom and current fashion. As we have seen from the stories in this book, these revolve not around technology, but people. For individuals and organisations that played such a remarkable role in advancing and applying technology, although these staff were excited by machines, they were also instinctively unwilling to over-rely on them. They could clearly see, in the case of Enigma, what happened to those who did.

This meant that the Bletchley pioneers eschewed the breathless utopianism of tech billionaires. This is partly because it was their job to see the downsides of technology; they were tackling adversaries who, after all, were using it to do considerable harm. And the key to defeating these enemies was partly to understand their technology, but primarily to understand the human thought processes and weaknesses that lay behind its use. In short, an organisation that produced brilliant technology was not over-confident in it, or naively utopian about what it could do.

If valuing people more than brilliant machines was the key, then finding suitable recruits – to perform tasks that were barely understood and generally regarded as impossible – could not follow convention and involved taking risks.

It meant searching for people who enjoyed games and puzzles, not as a relief from the stress of work, but because the work itself extended along a continuum of puzzle-solving and creative play. In pursuit of innovation and creativity, it meant prizing imagination over simple rationality. It also meant prioritising curiosity and curiousness over conventional expectations and behaviours. In short, odd people were welcome, absorbed and enabled. Such an approach was not driven by an ideological attachment to diversity for its own sake, but from a belief in the primacy of merit and competence, and an experience of what worked. By the same token, these organisations accepted people who had been left behind when the physically fit and conventionally employable were called up, those who were often the equivalent of the ones picked last in playground football.

Once assembled, this diverse bunch of individuals with a wide range of abilities and skills found themselves working within a culture that not only tolerated difference, but encouraged internal challenge and welcomed contemplation of the impossible. They were not necessarily fast thinkers, fast processors or articulate bluffers, and as we have noted, many of them would probably not have passed modern tech company recruitment tests for aptitude and problem-solving.

The constraints of secrecy in this culture ensured some balance. Recruits needed outside interests because work and home had to be rigorously separated. Even today, in an organisation dedicated to analysing communications, staff have to step out of the modern world of hyper-connectivity, where every moment of the day has to be reported in a stream of social media consciousness. Counter-intuitively, some detachment from the world is at the heart of this success in creativity and innovation.

And at work, humility was a more natural and useful quality than celebrity. If individuals were not seeking personal glory, they were free to pursue the ultimate team sport. Whether that was working across the disparate Huts of Bletchley Park or teaming up with Polish mathematicians and French spies, these people pioneered a culture of deep national and international collaboration. This reached its zenith in the 'special relationship' with the United States, a country that Britian had been spying on until shortly before the war. While that partnership was underpinned by a formal legal agreement, it was fundamentally based on shared democratic values, on a personal commitment between Churchill and Roosevelt, and on personal relationships between staff at all levels. That personal tradition of trust endures, and remains at the heart of GCHQ's success.

At the organisational level, the difference between individuals was reflected in structures. The wildly disparate components of Bletchley or GCHQ each had different requirements. The organisation needed to ensure that all of those who staffed them could flourish: not just academic mathematicians and linguists, but department-store assistants, bank clerks, engineers, industrial production-line operators, guards, radio experts, caterers and secretaries. Sixth-form leavers and Fellows of the Royal Society, apprentices and private sector partners – they all had varying cultures and distinct ways of doing things. And once assembled, no one could easily be moved on; there was no simple option of hire-and-fire. Problems had to be resolved, and each person needed to be enabled to perform.

The result was an experiment in living with dissonance, tribes in tension pursuing a single mission with a high degree of local autonomy and limited meddling from above. This was a flat structure where different cultures could emerge, appropriate to the workforce and the task.

The drudgery of leadership

These people and this organisation, if that is what it was, needed some form of leadership. I have said very little about leaders in this secret world of technology, except in passing. This is because they are not the most important part the story, in itself a conclusion that runs counter to modern wisdom and the celebrity status of modern tech bosses. But these people's primary function is to generate profit, not to create or engineer. Their innovation is commercial rather than technological, and their great achievements are generally the founding of business empires rather than coming up with technological breakthroughs.

In the world of Bletchley's secret innovation, few of the senior staff had received any training – or possessed much experience – in leadership. They tried out varying styles, from Milner-Barry's 'aimless wandering' to Turing's solution of allowing Hugh Alexander to do it for him. As long as it worked, no one minded. The much-maligned head of the whole show, Alastair Denniston, assembled the cast and later saw it as his job to protect all those involved from outside interference. His primary function, and Travis's after him, was to keep everyone focused on the mission, to enable them to excel and to remove blockages.

Denniston's job as leader was not to dictate solutions, and he knew in any case that he lacked the expertise to do so. In fact, many of the leaders were open about their limitations. Milner-Barry, the international chess master and one of the four, with Turing, who wrote to Churchill asking for urgent support, confessed to being 'almost innumerate'. Writing fifty years later, he said,

> to this day I could not claim that I fully understood how the machine worked, let alone what was involved in the problems of breaking and reading the Enigma cipher.

Roy Jenkins was posted to the new Colossus project and admitted that 'I never really understood what I was trying to do', a lesson he recalled when he later became minister of aviation.

Denniston's example is worth remembering. Traditional leaders find comfort in clearly demarcated structures and lines of control. Chaos and anarchy make them feel uncomfortable. The obvious danger is that they therefore mould the organisation not only to their own vision but also to their own limitations. The boundaries of their own imagination and understanding unconsciously define the potential extent of the organisation's development.

In an age of hyper-complex and inter-related problems, where machines will increasingly be our partners, these approaches to leadership look dated. The permanent disruption of technology and the discord we see in everything from culture and politics to health, education and our natural environment, call for different models if we are to innovate on the scale required.

As the first leader of GC&CS, Denniston collected the skills that he knew would be necessary and a range of other skills which he thought just might be. He could not be sure, but he took advice from people whose assessment of these needs he knew to be reliable and pursued a doggedly meritocratic approach. He was hiring for roles that had not been invented, in pursuit of a task whose attainment might not be possible. We have already seen that he spotted the importance of mathematicians in what was at the time a cryptologic world of languages; but his perception that machines and engineering would be valuable – in ways he did not understand – led him to throw engineers from the unfashionable General Post Office and British Tabulating Machine Company into the mix, a decision that turned out to be critical. Bletchley established a tradition of GCHQ staff setting their recruitment tests and taking part in selection, rather than delegating these to other branches of the civil service. Their job was to assess the whole person: what contribution could they make?

And in his un-showy way, Denniston took risks innovating. He and his successors supported leaders such as Max Newman in allowing competing approaches to problems to be run simultaneously. Even at a time of material shortages, scarce skills and constrained finance, Bletchley had a healthy approach to probability in innovation (hardly surprising, given that probability lay at the heart of codebreaking). Modern governments, by contrast, have often struggled in this area; they are so paralysed by the fear of being seen to waste money that they find it hard to fund risky innovation.

In pursuit of this innovation, these leaders also tolerated the collateral consequences of their recruitment, which outside observers regarded as extreme, and indulgent to the point of weakness. 'Eccentricity' was a given, neither disapproved of nor fetishised for its own sake. Dissent was not crushed and questioning was welcomed, because these were seen as fertile means to improvement and refinement. Because the problems were so new and so complex, it was self-evident that no single person had a monopoly on all the possible answers. An eighteen-year-old student or a young telephone engineer might well find a solution that had not occurred to their more distinguished bosses. In doing so, Bletchley Park discovered, long before its everyday application in high-performance sport, the importance of incremental advantage, or what are known as 'marginal gains'. Every slight improvement reduced the number of possibilities ranged against them and increased the probabilities of breaking Enigma.

In creating this multi-disciplinary soup of creativity, underpinned by a youthful, predominantly female workforce who believed that anything was possible, Bletchley's leadership imposed the minimum necessary controls and avoided rigid hierarchy. Flexibility and adaptation were given priority over structural coherence and tidiness. Where necessary, the hierarchical norms, particularly those of the military, were challenged. But conflict was never sought out for its own sake. Above all, the small centre absorbed outside demands, complaints and metrics, allowing the irregular 'honeycomb' to keep buzzing. The leadership was, of course, endlessly criticised by staff, who needed more resources and were frustrated by the pace of change.

And Bletchley was not always rollicking good fun. Retyping messages in code that made no sense – for ten hours at a stretch and marooned in the middle of nowhere – was neither romantic nor fulfilling for the young. There was grind and menial work and numerous irritations, not least for men of fighting age who would constantly be asked why they were not serving with the forces, yet could give no answer. But the staff, however brilliant, only experienced one side of the equation; the leadership spared them sight of Bletchley's pressures and organisational complexities, catalogued for posterity in staggering detail by the assiduous original 'dormouse' of Room 40, Nigel de Grey.

In short, this was a leadership that could live with tension without feeling compelled to resolve it immediately. In holding that tension, it resisted the natural tendency of any organisation to homogenise, standardise and dilute. Here were leaders who did not feel threatened by staff more expert and brilliant than they were, leaders who were self-effacing and verging on the reluctant. At their best, they embodied the paradox of this secret world. By understanding and prizing people above technology, they enabled the greatest technological advances imaginable.

Sources

Alongside the National Archives and those of Bletchley Park, GCHQ and the US National Security Agency have published excellent historical material in recent years.

There are many works on the subjects referred to in this book and I am indebted to them. A selection is listed here.

Abrutat, D., *Radio War: The Secret Espionage War of the Radio Security Service 1938–1946*, Fonthill Media, 2019

Agar, J., *The Government Machine: A Revolutionary History of the Computer*, MIT Press, 2019

Aldrich, R. J., Cormac, R. and Goodman, M. S., *Spying on the World: The Declassified Documents of the Joint Intelligence Committee, 1936–2013*, Edinburgh University Press, 2014

Alexander, C. O., *Book of Chess*, Hutchinson, 1973

Andrew, C. M., *The Secret World: A History of Intelligence*, Penguin Books, 2019

Baron-Cohen, S., *The Pattern Seekers: How Autism Drives Human Invention*, Basic Books, 2023

Batey, M., *Dilly: The Man Who Broke Enigmas*, Biteback Publishers, 2017

Beesly, P., *Room 40: British Naval Intelligence, 1914–1918*, Oxford University Press, 1984

Bennett, R. F., *Behind the Battle: Intelligence in the War with Germany, 1939–45*, Marble Hill, 2023

Benson, M., *Space Odyssey: Stanley Kubrick, Arthur C. Clarke, and the Making of a Masterpiece*, Simon & Schuster, 2019

Blum, A., *Tubes: Behind the Scenes at the Internet*, Penguin, 2013

Briggs, A., *Secret Days: Code-Breaking in Bletchley Park*, Frontline Books, 2015

Bright, C., *Submarine Telegraphs: Their History, Construction, and Working*, Cambridge University Press, 2014

Bright, E. B., *Life Story of the Late Sir Charles Tilston Bright: Civil Engineer With Which Is Incorporated the Story of the Atlantic Cable*, Forgotten Books, 2019

Budiansky, S., *Battle of Wits: The Complete Story of Codebreaking in World War II*, Touchstone, 2002

Calvocoressi, P., *Top Secret Ultra*, M. & M. Baldwin, 2001

Cassam, Q., *Conspiracy Theories*, Polity Press, 2021

Clark, R., *The Man Who Broke Purple*, Bloomsbury, 2012

Clayton, A., *The Enemy Is Listening*, Crécy Books, 1993

Cohen, M. N., *Lewis Carroll: A Biography*, Macmillan, 2015

Comer, T., 'Competing ineptitudes' and other articles on the Sigint Historian blog (https://siginthistorian.blogspot.com), 2022–23

Cookson, G., *The Cable*, Tempus, 2006

Copeland, B. J., *The Essential Turing: Seminal Writings in Computing, Logic, Philosophy, Artificial Intelligence, and Artificial Life plus The Secrets of Enigma*, Oxford University Press, 2013

Copeland, B. J., *The Turing Guide*, Oxford University Press, 2017

Copeland, B. J., *Colossus: The Secrets of Bletchley Park's Codebreaking Computers*, Oxford University Press, 2010

Corera, G., *Intercept: The Secret History of Computers and Spies*, Weidenfeld & Nicolson, 2016

Cox, P., *Spedan's Partnership: The Story of John Lewis and Waitrose*, Labatie Books, 2010

Danesi, M., *The Puzzle Instinct: The Meaning of Puzzles in Human Life*, Indiana University Press, 2002

Denniston, R. and Denniston, A. G., *Thirty Secret Years: A. G. Denniston's Work in Signals Intelligence, 1914–1944*, Polperro Heritage Press, 2007

D'Imperio, M. E., *The Voynich Manuscript: An Elegant Enigma*, Aegean Park Press, no date

Dooley, J. F., *The Gambler and the Scholars: Herbert Yardley, William & Elizebeth Friedman, and the Birth of Modern American Cryptology*, Springer, 2023

Du Sautoy, M., *The Creativity Code: How AI Is Learning to Write, Paint and Think*, 4th Estate, 2020

Eagleman, D., *Livewired*, Canongate, 2021

Erskine, R. and Smith, M., *Bletchley Park Codebreakers*, Biteback, 2011

Fagone, J., *The Woman Who Smashed Codes: A True Story of Love, Spies, and the Unlikely Heroine Who Outwitted America's Enemies*, Dey Street Books, 2018

Ferris, J., *Behind the Enigma*, Bloomsbury, 2020

Fitzgerald, P., *The Knox Brothers: Edmund 1881–1971, Dillwyn 1884–1943, Wilfred 1886–1950, Ronald 1888–1957*, 4th Estate, 2013

Friedman, W. F. and Friedman, E., *The Shakespearean Ciphers Examined: An Analysis of Cryptographic Systems Used as Evidence that Some Author Other than William Shakespeare Wrote the Plays Commonly Attributed to Him*, Cambridge University Press, 2010

Gambetta, D. and Hertog, S., *Engineers of Jihad: The Curious Connection Between Violent Extremism and Education*, Princeton University Press, 2018

The GCHQ Puzzle Book, Penguin, 2016

Gertner, J., *The Idea Factory: Bell Labs and the Great Age of American Innovation*, Penguin, 2018

Gilbert, M., *Churchill and the Jews*, Simon & Schuster, 2008

Goldsmid, F., *Telegraph and Travel*, Forgotten Books, 2023

Greenberg, J., *Alastair Denniston: Code-Breaking from Room 40 to Berkeley Street and the Birth of GCHQ*, Frontline Books, 2022

Greenberg, J., *Gordon Welchman*, Pen and Sword, 2022

Grey, Christopher, *Decoding Organization: Bletchley Park, Codebreaking and Organization Studies*, Cambridge University Press, 2013

Hafner, K. and Lyon, M., *Where Wizards Stay Up Late: The Origins of the Internet*, Simon & Schuster, 2006

Hannigan, R., 'Organising a Government for Cyber: The Creation of the UK's National Cyber Security Centre', Royal United Services Institute (RUSI), 27 February 2019

Hardy, G. H., *A Mathematician's Apology*, Origami Books, 2022

Herivel, J., *Herivelismus and the German Military Enigma*, M. & M. Baldwin, 2008

Hicks, M., *Programmed Inequality: How Britain Discarded Women Technologists and Lost Its Edge in Computing*, MIT Press, 2018

Hill, M., *Bletchley Park People: Churchill's 'Geese that Never Cackled'*, The History Press, 2004

Hinsley, F. H., *British Intelligence in the Second World War*, Volumes 1–5, HMSO, 1979–1990

Hinsley, F. H. and Stripp, A., *Codebreakers: The Inside Story of Bletchley Park*, Oxford University Press, 2011

Hodges, A., *Alan Turing: The Enigma*, Vintage, 1987

Howard, K., *Dear Code Breaker: The Letters of Margaret Rock (Bletchley Park Code Breaker) and John Rock (Parachute & Glider Forces Pioneer)*, Booktower Publishing, 2013

Hoy, H. C. and Mendelsohn, C. J., *40 O.B., Or, How the War Was Won*, Hutchinson, 1932

Jeffery, K., *The Secret History of MI6*, Penguin, 2010

Jennings, C., *The Third Reich Is Listening: Inside German Codebreaking 1939–45*, Osprey Publishing, 2019

Jones, R. V., *Most Secret War*, Michael Joseph, 2009

Kahn, D., *The Codebreakers: The Story of Secret Writing*, Scribner, 1996

Kahn, D., *Reader of Gentlemen's Mail*, Yale University Press, 2011

Kahn, D., *Seizing the Enigma*, Pen and Sword, 2022

Kenyon, D., *Bletchley Park and D-Day: The Untold Story of How the Battle for Normandy Was Won*, Yale University Press, 2022

Levy, S., *Crypto: Secrecy and Privacy in the New Code War*, Penguin, 2002

Lewin, R., *Ultra Goes to War: The Secret Story*, Pen and Sword Aviation, 2008

Lomas, D., 'Profiles in intelligence: an interview with Tony Comer', *Intelligence and National Security*, 38(1), 2023, pp. 1–15, https://doi.org/10.1080/02684527.2022.2090741

Macintyre, B., *Double Cross: The True Story of the D-Day Spies*, Bloomsbury Publishing, 2012

Madhavan, G., *Applied Minds: How Engineers Think*, W. W. Norton, 2016

Martin, K. M., *Cryptography: The Key to Digital Security, How It Works, And Why It Matters*, W. W. Norton, 2021

McKay, S., *Secret Listeners: How the Wartime Y Service Intercepted the Secret German Codes for Bletchley*, Aurum Press, 2012

Mundy, L., *Code Girls*, Little, Brown, 2018

Munson, R., *George Fabyan: the Tycoon Who Broke Ciphers, Ended Wars, Manipulated Sound, Built a Levitation Machine, and Organized the Modern Research Center*, Porter Books, 2013

Omand, D., *Principled Spying: The Ethics of Secret Intelligence*, Georgetown University Press, 2018

Page, G., *They Listened in Secret: More Memories of the Wrens*, George R. Reeve Ltd, 2003

Parrish, T., *The American Codebreakers: The U.S. Role in Ultra*, Scarborough House, 1991

Pepper, D., 'The business of SIGINT: The role of modern management in the transformation of GCHQ', *Public Policy and Administration*, 25(1), 2010, https://doi.org/10.1177/0952076709347080

Petzold, C., *Code*, Microsoft Press, 1999

Ratcliff, R. A., *Delusions of Intelligence: Enigma, Ultra, and the End of Secure Ciphers*, Cambridge University Press, 2008

Rid, T., *Rise of the Machines: The Lost History of Cybernetics*, Scribe, 2017

Roberts, J., *Lorenz: Breaking Hitler's Top Secret Code at Bletchley Park*, The History Press, 2018

Rosenheim, S., *The Cryptographic Imagination: Secret Writing from Edgar Poe to the Internet*, Johns Hopkins University Press, 1997

Rowlett, F. B., *The Story of Magic: Memoirs of an American Cryptologic Pioneer*, Aegean Park Press, 2002

Sebag-Montefiore, H., *Enigma: The Battle for the Code*, Orion, 2017

Sherman, D., *The First Americans: The 1941 US Codebreaking Mission to Bletchley Park*, Center for Cryptologic History, 2016

Sherman, W.H., 'How to Make Anything Signify Anything', *Cabinet*, Issue 40, 2011

Silberman, S. and Sacks, O., *Neurotribes: The Legacy of Autism and How to Think Smarter About People Who Think Differently*, Allen & Unwin, 2016

Singh, S., *The Code Book*, 4th Estate, 2000

Slimming, J., *Codebreaker Girls*, Pen and Sword Military, 2023

Smith, C., *The Hidden History of Bletchley Park: A Social and Organisational History, 1939–1945*, Palgrave Macmillan, 2017

Smith, G., Stephen, L. and Lee, S., *The Dictionary of National Biography*, Oxford University Press, 1998

Smith, M., *Station X*, Pan Macmillan, 2004

Smith, M., *The Secrets of Station X: How the Bletchley Park Codebreakers Helped Win the War*, Biteback, 2011

Smith, M., *Emperor's Codes: Bletchley Park's Role in Breaking Japan's Secret Ciphers*, Biteback, 2022

Snow, C. P., *Science and Government*, Harvard University Press, 1961

Sonni, J. and Goodman, R., *A Mind at Play: How Claude Shannon Invented the Information Age*, Simon & Schuster, 2018

Stripp, A., *Codebreaker in the Far East*, Oxford University Press, 2009

Sugarman, M. and Kushner, M. A., *Bletchley Park: The Jewish Contribution*, Michael A. Kushner, 2019

Syed, M., *Black Box Thinking: The Surprising Truth About Success (and Why Some People Never Learn from Their Mistakes)*, John Murray, 2015

Turing, D., *Prof: Alan Turing Decoded*, The History Press, 2016

Turing, D., *Codebreakers of Bletchley Park: The Secret Intelligence Station that Helped Defeat the Nazis*, Arcturus, 2020

Turing, S., *Alan M. Turing*, Cambridge University Press, 2015

Ui Chionna, J., *Queen of Codes: The Secret Life of Emily Anderson, Britain's Greatest Female Code Breaker*, Headline, 2024

Welchman, G., *The Hut Six Story: Breaking the Enigma Codes*, M. & M. Baldwin, 1997

West, T. G. and Sacks, O., *In the Mind's Eye: Creative Visual Thinkers, Gifted Dyslexics, and the Rise of Visual Technologies*, Prometheus Books, 2020

wikipedia.org, a particularly strong source for cryptography

Winkler, J. R., *Nexus: Strategic Communications and American Security in World War I*, Harvard University Press, 2013

Yardley, H. O., *The Education of a Poker Player*, High Stakes, 2005

Yardley, H. O., *American Black Chamber*, ISHI Press, 2016

Zacharias, E. M., *Secret Missions: The Story of an Intelligence Officer*, Naval Institute Press, 2014

Index